AN EXPERIMENT WITH TIME

By the same Author

★

The Serial Universe
Nothing Dies
Intrusions?

AN EXPERIMENT WITH TIME

by

J. W. DUNNE

FABER AND FABER LIMITED
24 Russell Square
London

First published in mcmxxvii
First published in this edition mcmlviii
by Faber and Faber Limited
24 Russell Square, London, W.C.1
Reprinted mcmlx and mcmlxiv
Printed in Great Britain by
Purnell and Sons, Limited
Paulton (Somerset) and London

INTRODUCTION
TO THE THIRD EDITION

An Experiment with Time was published first in March of 1927. It has passed through two editions and a reprint, without any substantial alteration. For this (third) edition I have thought it advisable to overhaul the book from beginning to end. I have inserted about eighty pages of new matter (including a new chapter, XI*a*), and I have done my best to simplify still further the arguments in the analytical chapters. The most important addition, however, is Appendix III, which deals with a new method of assessing the value of the evidence obtained. This amounts, in effect, to a new experiment of very great potency.

The general reader will find that the book demands from him no previous knowledge of science, mathematics, philosophy, or psychology. It is considerably easier to understand than are, say, the rules of Contract Bridge. The exception is the remainder of this Introduction. That has been written entirely for specialists, and is in no way a sample of what is to come.

* * *

Multidimensional worlds of the kind beloved by mystics, and dating back to the days of the Indian philosopher Patañjali, have never appealed to me. To introduce a new dimension as a mere hypothesis (*i.e.*, without logical compulsion) is the most extravagant proceeding possible. It could be justified only by the necessity of explaining some insistent fact which would appear, on any other hypothesis, miraculous. And a new and still more marvellous miracle would need to be discovered before we could venture to consider the possibility of yet another dimension. Even then the major difficulty would remain to be overcome. For why should the, say, five-dimensional observer of a five-dimensional world perceive that world as extended in only three dimensions?

The universe which develops as a consequence of what is known to philosophers as the 'Infinite Regress' is entirely free from the foregoing objections.

This 'Infinite Regress', I may explain to the uninitiated, is a curious logical development which appears immediately one begins to study 'self-consciousness' or 'will' or 'time'. A self-conscious person is one 'who knows that he knows'; a willer is one who, after all the motives which determine choice have been taken into account, can choose between those *motives*; and time is——but this book is about that.

The usual philosophic method of dealing with any regress is to dismiss it, with the utmost promptitude, as something 'full of contradictions and obscurities'. Now, at the outset of my own perplexing experiences, I supposed that this attitude was justified. But the glaring regress in the notion of 'time' was a thing which had intrigued me since I was a child of nine (I had asked my nurse about it). The problem had recurred to me at intervals as I grew older. I had troubles enough without this one, and I wanted it out of the way. Finally, I set to work to discover *what* were the contradictions and *where* were the obscurities. I spent two years hunting for the supposed fallacy. None, I think, can have subjected this regress to a fiercer, more varied or more persistent attack. These assaults, to my great surprise, failed. Slowly and reluctantly I acknowledged defeat. And, at the end, I found myself confronted with the astonishing facts that the regressions of 'consciousness', 'will' and 'time' were perfectly logical, perfectly valid, and the true foundations of all epistemology.

It was not, however, until years later that it dawned upon me wherein lay the full significance of any regress. A regress is merely a mathematical series. And that is merely the expression of some relation. But the relation thus expressed is one which does not become apparent until one has studied the *second term* of the series concerned. Now, the second term of the regress of time brings to light relations of considerable importance to mankind. It is the existence of these relations that the regress asserts. But the information thus disguised is entirely lost if we confine our study to the opening term alone. Yet that is what mankind has been doing.

As soon as I realized this I sat down and wrote the book. It contains the first analysis of the Time Regress ever completed. Incidentally, it contains the first scientific argument for human immortality. This, I may say, was entirely unexpected. Indeed, for a large part of the time that I was working, I believed that I was taking away man's last hope of survival in a greater world.

J. W. DUNNE

March 15*th*, 1934.

EXTRACT FROM A NOTE ON THE SECOND EDITION

It has been rather surprising to discover how many persons there are who, while willing to concede that we habitually observe events before they occur, suppose that such prevision may be treated as a minor logical difficulty, to be met by some trifling readjustment in one or another of our sciences or by the addition of a dash of transcendentalism to our metaphysics. It may well be emphasized that no tinkering or doctoring of that kind could avail in the smallest degree. If prevision be a fact, it is a fact which destroys absolutely the entire basis of all our past opinions of the universe. Bear in mind, for example, that the foreseen event may be avoided. What, then, is its structure?

I would suggest that we are lucky, on the whole, to be able to replace our vanished foundations by a system so simple as the 'serialism' described in this book.

Anyone who hopes to discover an explanation even simpler would be well advised to examine his own statement of the difficulty to be faced—viz., that we 'observe events before they occur'. Let him ask himself to what time-order does that word '*before*' refer. Certainly not to the primary time-order in which the occurring events are arranged! He may see then that his statement (and every expression of his problem must bear that same general form) is in itself a direct assertion that Time is serial.

If Time be serial, the universe as described in terms of Time must be serial, and the descriptions, to be accurate, must be similarly serial—as suggested in Chapter XXV. If that be the case, the sooner we begin to recast physics and psychology on such lines, the sooner may we hope to reckon with our present discontinuities and set out upon a new and sounder pathway to knowledge.

J. W. DUNNE

CONTENTS

PART I

DEFINITIONS

CHAPTER I

It might, perhaps, be advisable to say here—since the reader may have been glancing ahead—that this is not a book about 'occultism', and not a book about what is called 'psycho-analysis'.

It is merely the account of an extremely cautious reconnaissance in a rather novel direction—an account presented in the customary form of a narrative of the actual proceedings concerned, coupled with a statement of the theoretical considerations believed to be involved —and the dramatic, seemingly *bizarre* character of the early part of the story need occasion the reader no misgivings. He will readily understand that the task which had to be accomplished at that stage was the 'isolating' (to borrow a term from the chemists) of a single, basic fact from an accumulation of misleading material. Any account of any such process of separation must contain, of course, some description of the stuff from which the separation was effected. And such stuff very often is, and in this case very largely was—rubbish.

* * *

There does not appear to be anything in these pages that anyone is likely to find difficult to follow, provided that he avoids, in Chapters XVII, XIX, XXI, XXIII, XXIV and XXVI, those occasional paragraphs enclosed within square brackets which have been written more particularly for specialists. And Part V may require reading twice. But there are a few commonplace semi-technical expressions which will crop up now and again; and it is always possible that other people may be accustomed to attach to these words meanings rather different from those which the present writer is hoping to convey. Any such misunderstanding would result, obviously, in our being at cross-purposes throughout the greater part of the book. Hence it might be advisable for us to come to some sort of rough preliminary agreement, not as to how these terms ought rightly to be employed, but as to what they are to be regarded as *meant* to mean in this particular volume. By so doing we shall, at any rate, avoid that worst of all irritations to a reader—a text repeatedly interrupted by references to footnote or glossary.

That the agreement will be entirely one-sided will make it all the easier to achieve.

CHAPTER II

Briefly, then:

Let us suppose that you are entertaining a visitor from some country where the inhabitants are all born blind; and that you are trying to make your guest understand what you mean by 'seeing'. You discover, we will further assume, that the pair of you have, fortunately, this much in common: You are both thoroughly conversant with the meanings of all the technical expressions employed in the physical sciences.

Using this ground of mutual understanding, you endeavour to explain your point. You describe how, in that little camera which we call the 'eye', certain electromagnetic waves radiating from a distant object are focused on to the retina, and there produce physical changes over the area affected; how these changes are associated with currents of 'nervous energy' in the criss-cross of nerves leading to the brain-centres, and how molecular or atomic changes at those centres suffice to provide the 'seer' with a registration of the distant object's outline.

All this your visitor could appreciate perfectly.

Now, the point to be noticed is this. Here is a piece of knowledge concerning which the blind man had no previous conception. It is knowledge which he cannot, as you can, acquire for himself by the ordinary process of personal experiment. In substitution, you have offered him a *description*, framed in the language of physical science. And that substitute has served the purpose of conveying the knowledge in question from yourself to him.

But in 'seeing' there is, of course, a great deal more than mere registration of outline. There is, for example—Colour.

So you continue somewhat on the following lines. That which we call a 'red' flame sets up electro-magnetic waves of a certain *length*: a 'blue' flame sets up waves exactly similar save only that they differ slightly in this matter of length. The visual organs are so constituted that they sort out waves showing such disparity in length, and this in such a way that these differences are finally registered by corresponding differences in those physical changes which occur at the brain centres.

From the point of view of your blind guest, this description, also, would be entirely satisfactory. He could now understand perfectly how it is that a physical brain is able to register wave-length difference. And, if you were content to leave it at that, he would depart gratefully convinced that the language of physics had again proved equal to the task, and that your description in physical terms had equipped him with a knowledge of, for instance, what other people call 'red' as complete in every respect as that which they themselves possess.

But this supposition of his would be absurd. For concerning the existence of one very remarkable characteristic of red he would still, obviously, know nothing whatsoever. And that characteristic (possibly the most puzzling, and certainly the most obtrusive of them all) is—its *redness*.

Redness? Yes. Without bothering about whether redness be a thing or a quality or an illusion or anything else, there is no escaping the fact (1) that it is a characteristic of red of which you and all seeing people are very strongly aware, nor the further fact (2) that your visitor, so far, would have not the faintest shadow of an idea that you or others experience anything of the kind, or, indeed, that there could exist anything of the kind to be experienced. If, then, you intend to complete your self-imposed task of bringing his knowledge on the subject of 'seeing' up to the same level as your own, there remains yet another step before you.

Realizing this, you mentally glance down your list of physical expressions, and—a moment's inspection is enough to show you that, for the purpose of conveying to your blind guest a description of *redness*, there is not a single one of these expressions which is of the slightest use whatsoever.

You might talk to him of particles (lumps—centres of inertia), and describe these as oscillating, spinning, circling, colliding, and rebounding in any kind of complicated dance you cared to imagine. But in all that there would be nothing to introduce the notion of *redness*. You might speak of waves—big waves, little waves, long waves, and short waves. But the idea of *redness* would still remain unborn. You might hark back to the older physics, and descant upon forces (attractions and repulsions), magnetic, electrical, and gravitational; or you might plunge forward into the newer physics, and discourse of non-Euclidean space and Gaussian co-ordinates. And you might hold forth on such lines until exhaustion supervened, while the blind man nodded and smiled appreciation; but it is obvious that, at the end of it all, he would have no more suspicion of what it is that (as Ward puts it) 'you immediately experience when you look at a field poppy' than he had at the outset.

Physical description cannot here provide the information which experience could have given.

Now, redness may not be a thing—but it is very certainly a *fact*. Look around you. It is one of the most staring facts in existence. It challenges you everywhere, demanding, clamouring to be accounted for. *And the language of physics is fundamentally unadapted to the task of rendering that account.*

It is obvious that dubbing redness an 'illusion' would not

help the physicist. For how could physics set about describing or accounting for the entry of the element of *redness* into that illusion? The universe pictured by physics is a colourless universe, and in that universe all brain-happenings, including 'illusions', are colourless things. It is the intrusion of Colour into that picture, whether as an illusion or under any other title, which requires to be explained.

Once you have thoroughly realized that redness is something beyond a complex of positions, a complex of motions, a complex of stresses, or a mathematical formula, you will have little difficulty in perceiving that Colour is not the only fact of this kind. If your hypothetical visitor were deaf, instead of blind, you could never, by giving him books of physics to read, arouse in him even the beginning of a suspicion regarding the nature of 'Sound', as *heard*. Now, Sound, as heard, is a fact. (Put down this book and listen.) But in the world described by physics there is no such fact to be found. All that physics can show us is an alteration in the *positional arrangement* of the brain-particles, or alterations in the *tensions* acting upon those particles. And in no catalogue of the magnitudes and directions of such changes could there be anything to suggest that there exists anywhere in the universe a phenomenon such as that which you directly experience when a bell tolls. In fact, just as physics cannot deal with the element of redness in 'red', so is it inherently unable to account for the intrusion of that clear bell-note into a universe which it can picture only as an animated diagram of groupings, pushings, and pullings.

But if, in such a diagram, there can be nothing of either Colour or Sound, is it likely to be of any use our hunting therein for phenomena like 'Taste' and 'Smell'? The utmost that we could hope to find would be those movements of the brain-particles which *accompany* the experiences in question; or, possibly, some day, the transference equations relating to some hitherto unsuspected circuit of energy. Your hypothetical visitor and yourself might each possess the fullest possible knowledge of these brain-disturbances, the most complete acquaintance with such energetic equations as may still remain to be written; but, if *you* could actually taste and smell, and he could not, it is incontrovertible that your knowledge of each of these phenomena would include something quite unknown to, and, indeed, quite unimaginable by, him.

Now, when we say of any occurrence that it is 'physical', we mean thereby that it is potentially describable in physical terms. (Otherwise the expression would be wholly meaningless.) So it is perfectly correct to state that, in every happening with which our sensory nerves are associated, we find, *after* we have abstracted therefrom every known or imaginable physical component, certain categorically nonphysical *residua.*

But these remnants are the most obtrusive things in our universe. So obtrusive that, aided and abetted by our trick of imagining them as situated at our outer nerve-endings, or as extending beyond those endings into outer Space, they produce the effect of a vast external world of flaming lights and colours, pungent scents, and clamorous, tumultuous sounds. Collectively, they bulk into a most amazing tempest of sharply-differentiated phenomena. And it is a tempest which remains to be considered *after* physics has completed its say.

Physics.

Nor is this last a matter for wonderment. For the ideal object of physics is to seek out, isolate, and describe such elements in Nature as may be credited with an existence independent of the existence of any human observer. Physics is, thus, a science which has been expressly designed to study, not the universe, but the things which would supposedly remain in that universe if we were to abstract therefrom every effect of a purely sensory character. From the very outset, then, it renounces all interest in such matters as those colours, sounds, etc., of which we are directly aware,—matters essentially dependent upon the presence of a human observer, and non-existent in his absence,—and it limits itself to a language and a set of conceptions serviceable only for the description of facts pertaining to its own restricted province.

Psychology and Psychical.

But, as scientific investigators of the situation in which we find ourselves, we cannot, of course, neglect to study a mass of phenomena so large and so obtrusive as to constitute, to first appearance, the whole of the world we know. Consequently, a separate science has gradually arisen which endeavours to deal with these and other of the rather bulky leavings of physics. This science is called '*Psychology*', and the facts with which it deals are dubbed 'mental', or, more commonly, 'Psychical'.

CHAPTER III

No, although it is scientifically indisputable that the brain, regarded as a purely physical piece of mechanism, cannot create, unassisted and out of nothingness, any of those vivid psychical appearances we call 'colour', 'sound', 'taste', etc., it may be taken as experimentally established that these phenomena do not come into existence unless accompanied by some stimulation of the corresponding sense organs. Moreover, they vary in character according to the character of the sense organ involved: lights and colours accompany activities of the optic nerves; sounds are associated with the existence of ears; tastes with palates. The psychical phenomena are different where the sensory organizations are different. Colour experiences in man range from violet to deep red, according to the wave-lengths of the electro-magnetic rays impinging upon the eye. If that wave-length be further slightly increased, the associated psychical experience is one of heat alone. But we know that, with a very little modification of the sensitive optical elements involved, those heat experiences would be accompanied by experiences of a visible infra-red colour.

Thus, the physical brain, though it cannot create such sensory appearances, is a prime factor in their *characterization*, and, for that reason, an important factor in whatever process it may be that causes them to appear.

The situation, thus far, is usually summed up in the cautious statement that these particular kinds of psychical phenomena, on the one hand, and their corresponding sense-organ stimulations, on the other, invariably accompany one another, or run, so to say, on parallel tracks in

Time. This, be it noted, is never advanced as an 'explanation': it is merely supposed to be a simple way in which the facts can be announced without dragging in the various metaphysical creeds favoured by the various announcers.

Psychoneural Parallelism.

The assumption that this 'parallelism' of psychical and neural (nervous) events extends to *all* observable thought-experience—that there is no observable psychical activity without some corresponding activity of brain—is called '*Psychoneural Parallelism*'; the activity in either class being referred to as the 'correlate' of that in the other.

The accumulated evidence in favour of this view is practically overwhelming. Hard thinking induces brain fatigue; drugs which poison the brain interfere with our reasoning processes; brain deterioration affects our ability to form new memories. Above all, 'concussion' of the brain appears to destroy all memory of the events which immediately *preceded* the accident—indeed, it is by the failure of the patient to remember what led up to that accident that the physician diagnoses concussion. This provides us with almost indisputable evidence that the means of remembering are 'brain-traces' which *require a little time for their assured establishment.*

That such brain-traces (insulated paths formed by the passage of nervous currents) do, in fact, exist, is very probable; and moreover, it has been shown that the greater the ability of the individual to perform associative thinking, the more numerous and the more complex in their ramifications are the brain paths in question.

Observer.

We have now arrived within introductory range of that very meek-spirited creature known to modern science as the '*Observer*'. It is a permanent obstacle in the path of our search for external reality that we can never entirely get rid of this individual. Picture the universe how we may, the picture remains of our making. On the other hand, it is, probably, equally true that, paint the picture how we will, we have to do it with the paints provided. But there is no reason why either of these limitations should invalidate the result regarded as a 'map by which we may safely set our course. Moreover we can test it in that respect; and experience has shown that, thus tested, it proves reliable. Therein lies the justification of our search for knowledge.

The general procedure in every science is to begin by the accurate tabulating of differences in what is observed. If we subsequently discover that these differences are due to the character or actions of the observer, we can note that such is the explanation of the difference and draft our science accordingly; but that addition to our knowledge does not invalidate our previous analysis of the differences as observed.

All sciences deal only with a *standard* observer, unless the contrary is explicitly stated; and psychology is no exception to this rule. Its observer is assumed to be any normally constituted individual.

Now, it must be admitted that the tenets of psychoneural parallelism are not very encouraging to this 'observer'. For they suggest that, when the brain-workings come to an end, the psychical phenomena cease likewise from troubling. Moreover, the scientific procedure of pushing the observer as far back as possible—so as to get as much as possible of the picture into the category of that which is observed—tends to reduce him to the level of a helpless onlooker with no more capacity for interference than has a member of a cinema audience the ability to alter the course of the story developing before him on the screen.

Nor is there much more comfort to be obtained from a study of the various metaphysical interpretations (none of them offer an *explanation*) of this parallelism of Mind and Body. Idealist and Realist may dispute hotly as to precisely how far the observer colours, so to say, the phenomena which he observes; but decisions arrived at in that respect need not suggest that he has any power of changing either the colouring he confers or the thing perceived as thus coloured—much less the ability to continue observing when there is no longer any brain activity to be observed.

Animism.

In this connection, however, we must recognize the existence of a small but very vigorous group of philosophers known as '*Animists*'. In this twentieth century the leading exponent of Animism is indubitably Professor William McDougall, whose book, *Body and Mind*, sets out the arguments for and against the theory with scrupulous fairness. Indeed, I cannot call to mind anyone who has stated the case against Animism with such devastating force.

Animism holds that the observer is anything but a nonentity. He is no 'conscious automaton'. He may, indeed, stand right outside the pictured universe; but he is a 'soul', with powers of intervention which enable him to alter the course of observed events—a mind which not only reads the brain, but which employs it as a tool. Much as the owner of an automatic piano may either listen to its playing or play on it himself.

The inference is that this observer can survive the destruction of that brain which he observes. As for his intervention, there is no insuperable objection to that from the physical side. McDougall quotes and suggests various ways in which intervention could be effected without adding to or subtracting from the amount of energy in the nervous system.

The man-in-the-street is always at a loss to understand why the great majority of men of science are so coldly opposed to the ideal of a 'soul'. The religious man in particular cannot comprehend why his arguments should arouse not merely opposition, but bitter contempt. Yet the reason is not far to seek. It is not that the idea is attributed to man's inordinate conceit (though this is sometimes done by the unreflecting); for, all said and done, a navvy who can walk into a public-house and order a pot of beer is an infinitely more wonderful thing than is the biggest lump of cooling mud that ever swam in the skies. But there can be no reasonable doubt that the idea of a soul must have first arisen in the mind of primitive man as the result of observation of his dreams. Ignorant as he was, he could have come to no other conclusion but that, in dreams, he left his sleeping body in one universe and went wandering off into another. It is considered that, but for that savage, the idea of such a thing as a 'soul' would never have even occurred to mankind; so that arguments subsequently introduced to bolster up a case thus *tainted at its source* can have no claim to anyone's serious attention.

CHAPTER IV

Presentations.

Psychology must begin, then, by describing observed *appearances* (the literal translation of the word 'phenomena') without any prejudging of the issue as to what is the cause of these. So, though it may speak of such phenomena as if they were things, it must not be regarded as asserting that they are, at bottom, anything more than effects associated with brain-workings. It leaves, at the outset, that question open.

Field of Presentation.

All such phenomena it styles '*Presentations*', and it regards them as located within the individual's private '*Field of Presentation*'. (We shall employ this term in preference to the commoner 'Field of Consciousness', which is insufficiently definite.) This field of presentation contains, at any given instant of Time, all the phenomena which happen to be offered for possible observation. Let us take a concrete example of what that means. You are now reading this book, and your field of presentation contains the visual phenomena connected with the printed letters of the word you are regarding. It contains also, at the same instant, the visual phenomenon pertaining to the little numeral at the bottom of the page. This you 'failed to notice'; but the numeral in question was, clearly, inside the area covered by your vision—it was affecting your brain *via* the eye, its psychical 'correlate' was being offered to your attention. And that statement holds good for a host of other visual phenomena. On reflection, you will also agree that the field must have then contained—presented to attention but left 'unnoticed' —certain muscular sensations such as pressures against your body, quite a number of sounds, and the pleasant feeling produced by the air flowing into your lungs as you breathed.

Attention.

It would be unsafe to say that these comparatively unnoticed phenomena were not being consciously observed. When you are watching a fall of snow, observation may be concentrated upon a single floating flake; but that does not mean that you fail to perceive the remainder. Were these to vanish, leaving the single flake in the air, their disappearance would instantly distract your startled attention from the object of your previous preoccupation. When listening to the playing of an orchestra, you do not need to cease following the music in order to be aware that the irritating person in the seat ahead has stopped beating time with his programme. As a general rule, however, observation seems to be definitely centred upon one or another specific part of the crowd of presentations—though we have no physical evidence to show that this is anything more than a matter of habit. Observation thus centred is called '*Attention*'. It is usual to speak of the part of the field centred upon as being in the '*Focus of Attention*'; and it is a matter of common knowledge that, at and around this 'focus', attention may be concentrated in greater or less degree of intensity.[1]

In Physiology (the science which deals with the brain as a physical organism) the field of presentation would

[1] The reader should distinguish between the focusing of attention and the performing of body movements to assist observation. With his eyes focused upon a dull object before him, he can focus *attention* upon a brighter object in his field of vision. And he can, subsequently, shift the focus of vision to the brighter object.

be merely the particular part of this cerebrum which happens to be, at that moment, in the state of activity associated with the production of psychical phenomena. And the focus of attention would be simply that particular brain path which the maximum current of nervous energy happened to be following. One would be apt to suppose, off-hand, that this maximum flow would be produced by whatever happened to be the greatest sensory stimulation; but such could not be the rule. The hungry man, coming to the luncheon table, has his attention focused, not upon the brightness of the shining silver, but upon the far duller sensory stimulation of the well-browned mutton chop. Attention, therefore, may be either attracted from without the organism or directed from within. If we were to attribute such directing to the ultimate observer, we should be admitting him to the status of a full-blown 'animus' with powers of intervention. For, as every schoolboy knows, the concentrating of attention has a very marked effect in the formation of memories. But the physiologist would argue that we have no right to regard this internal directing of attention as originating in anything beyond the purely mechanical internal condition of the brain.

Now, the field of presentation at any given moment may contain a great many observable phenomena besides those sensory appearances which we have been considering. It may contain, for example, '*Memory Images*'.

What sort of a phenomenon is a 'memory image'?

Impressions.

Presentations may be divided into two sharply differing classes. The first of these comprises all phenomena which appear to the observer as directly attributable to the action of his outer sensory organs or nerve endings. That they are truly associated with the activities of such surface machinery is evident from the fact that movement of, or external interference with, the organs or nerve endings in question results in an alteration of the character of the phenomena observed, and from the equally significant fact that, in the absence of such movements or interferences, the phenomena remain unaltered and *unescapable*. They cannot, in popular parlance, be 'willed away'. Such phenomena are styled '*Impressions*'.

Images.

But now, picture to yourself a room which you remember. There is no doubt that what you are observing is a *visual* presentation—a mental *picture*. The process is not one of saying to yourself: 'Let me see: there was a sofa in that corner, and a piano in the other, and the colour of the carpet was such-and-such.' Rather does the whole of what you remember come before your eyes in the form of a simultaneous vision. If, however, you want to make absolutely certain that such visual pictures are not things which you deliberately manufacture from a catalogue of verbally remembered detail, you may try the following experiment. Look carefully at a painting of a landscape; then, after half an hour, try to re-visualize what you saw. You will find that you can re-observe much of the exact colouring of the original impressions—the peculiar olives and browns and greys—even though many of these colours were quite beyond your powers of artistic analysis, let alone verbal description. So you must be observing, as an 'image', an arrangement of colours similar to those which you saw as impressions.

Reality Tone.

There is a difference between an impression and its related image which it has puzzled every psychologist to describe. It lies in the presence or absence of what is sometimes called 'sensory vividness', but what, I think, would be better referred to as '*Reality Tone*'. As compared with a room which you can see with your eyes, the room you are remembering seems unreal, yet real enough to be recognizable as a *visual*, and not, say, an *aural* image. Again, strike the rim of a wine glass, and listen to the sound as it dies away. It grows fainter and fainter till it vanishes; but to the last (as Ward points out) it retains its reality tone. After it has entirely disappeared, you can remember what it sounded like just before it died away. That memory is recognizable as a memory of sound—an aural image. It has all the tonal qualities of the original faint impression; but it lacks the appearance of reality.

Again, compare the true memory image with the phenomenon commonly called an 'after-impression'. The latter may be easily observed. If you stare hard for sixty seconds at a brilliant red lampshade, and then look up at the ceiling, you will see, after a moment or so, a patch of *green*, shaped in outline like the lampshade. This phenomenon is dim, exhibits little, if any, detail within its boundaries, is of the opposite (complementary) colour to the original impression, and lacks all perspective—seeming to be flat all over. It possesses, however, reality tone, and is clearly an impression. It moves as you move your eyes. But, while actually watching this green patch floating before you, you can observe a true memory image of the original impression of the lampshade. It is of the original red colour, exhibits much internal detail, and appears to be three-dimensional—*i.e.*, to possess the depth apparent to binocular vision.

Five minutes later, when all trace of the green after-impression has vanished, you can observe at will clear memory images of either red lampshade or green patch.

It may be noted, then, that images are phenomena quite distinct from mere fading impressions.

CHAPTER V

Memory-Train.

Now, when you are trying to recall a *succession* of observed impressions, the images pertaining to these are observed as if they were actually arranged in an order corresponding to the order in which the original impressions were received. This supposed arrangement is called, as everybody knows, the '*Memory-Train*', and it is noticeable that the process of remembering events in the order in which they occurred is one which involves sometimes a very considerable mental effort. But if you are merely allowing your mind to wander—as in a day-dream—without knowingly aiming at any definite goal, the set of images which is then observed appears to be arranged in a sequence which has little correspondence with any previous observed succession of events.

Train of Ideas.

This curious succession of images is called the '*Train of Ideas*', and it is possibly a very significant fact that the simple, undirected following of a train of ideas appears to entail no mental effort or fatigue whatsoever.

Almost everybody has, at one time or another, amused himself by retracing the train of ideas which has led him, without any conscious aim on his part, to think of, or remember, a certain thing. 'I saw this', he will say, 'and that made me remember so-and-so; and that made me think of such-and-such.' And so on. Here, however, is a specific example.

It is now evening, and in front of me stands a teacup with a chequered black and white bordering. The sight of this (an impression) 'brings up' a memory image of the chequered oil-cloth floor-covering which, this morning, I was using as material for an experiment in obtaining after-impressions. Now, at the time of making that experiment I was thinking of Ward's description of these phenomena in the *Encyclopædia Britannica*; and the next image to appear before me is an image of the red volume in question (mine is the small-print edition). Following that, there appears an image of an open page in the volume, and a very vivid image of the sensation of eye-strain involved in its reading. That 'brings up' an image of the reading-glass I sometimes use. That 'brings up' the image of the lens I borrowed in a fishing-tackle shop yesterday morning in order to examine some trout flies I was buying. That 'brings up' the image of the friend for whom I had purchased those flies, as he stood when asking me to do so. And that 'brings up' the pleasing image of the two-and-a-half pound trout I annexed from that friend's water two days ago. Thus, starting with a teacup, I arrive at a trout.

Now, examination of the nature of a train of ideas brings to light the following facts.

Generic Images.

When a number of partly similar impressions have been attended to at different times, there is observable, besides the several memory images pertaining to those several impressions, a vague, general image comprising nothing beyond the key elements which are common to all those separate images. For example, the images of the hundreds of tobacco pipes which I have seen, smoked, and handled, all contain a common element which is now apparent to me as an ill-defined image of 'pipe' in general. It presents all the essential characteristics which serve to distinguish a pipe from any other article such as, say, an umbrella.

Such characteristics are: hollow bowl, tubular stem—in short, an appearance of utility for the purpose of smoking. But this indefinite image does not exhibit any indication of specific colour or precise dimensions. It seems, however, to be the nucleus of all the definite images of *particular* pipes to be found in my mental equipment; for, if attention be directed to it, there will quickly become observable the image of sometimes one and sometimes another of such particular pipes.

These vague, almost formless general images are called '*Generic Images*', and they appear to be analogous to a central knot to which the specific, definite images are in the relation of radiating threads.

Associational Network.

It is obvious that many of these threads—these definite images—may be radiating also from another generic image. A definite image of a particular wooden pipe-bowl may pertain, on one side to the generic image 'pipe', and, on another, to the generic image I call 'grained wood'. That generic image may have, as another of its components, a definite image of a polished walnut table, which image, again, may also be a radiating thread pertaining to the generic image 'furniture'. A thread from 'furniture'—say, the image of a particular suite seen in a shop window—may be the link with the generic image 'antiquities'. So far, then, we are confronted with something analogous to a network of knots (generic images) and radiating threads (definite images) along the meshes of which attention may be led without conscious effort on the part of the observer. Ideas linked together in this manner graphically analogous to a network of knots and threads are said to be 'associated'. Hence we may refer to the structure in question as the '*Associational Network*'.

It is commonly assumed that association is of two kinds: association by *similarity*, as when one event recalls a similar event which may have happened long ago; and association by *contiguity*, which means that, when two events have occurred in close succession, the recalling of one leads to the recollection of the other.

To the physiologist the associational network is simply the network of brain-paths, the 'knots' being regions—or patterns—therein, and the 'connecting threads' being paths which pertain to more than one such region—or pattern. All the phenomena of association seem to be adequately accounted for on that supposition; and on no other theory, so far as I can see, is it possible to account for association by '*similarity*' at all.

In the absence of any other guidance, the path taken by the train of ideas seems to be conditioned very largely by the factor of *freshness* in the images. Other things being equal, an image which has been recently established makes a stronger bid for the wandering attention than does one which has long been neglected. The reader will notice that, in the example of a train of ideas given a little way back (the one which began with a teacup), all the images related to experiences which had recently occurred. For example, the black and white chequering of the teacup led me, not to chess, which is a very obtrusive generic image of mine, but to the piece of linoleum I had seen that morning. Physiologically, this would mean that brain-paths which have been recently traversed offer a better passage to the currents of nervous energy than do those which have been allowed to fall into disuse.

The supposed 'memory-train' does not appear to be anything more than a particular pathway through the associational network, the pathway which happens to have been thus recently traversed. If you try to trace a 'memory-train' back for more than a little way, you find that the path has ceased to be clearly marked out: the images do not come up in a steadily correct sequence of, so to say, their own accord. You have to help the memory out by *reasoning* as to which event must have happened next—and sometimes you reason wrongly.

Dreams.

Dreams, like many other mental phenomena, are composed largely of images supplied by an associational network. But they differ from mind-wandering in several important respects. In the latter form of activity reason is nearly always partially at work to determine the course to be followed along the network. But in dreams this guidance seems to be largely lacking, and the dream images present themselves as real—though curiously unstable—episodes in a personal adventure story of an only partially reasonable character.

Integration.

Association between the dream-images is sometimes clear enough; but, as a general rule, such association takes the curious form known as an '*Integration*'. By this word we shall mean: 'A combination of associated images in which the composing elements are qualitatively distinguishable.' (This definition is from Baldwin's *Dictionary of Philosophy and Psychology*.) For example, the image of a pink dress seen in a shop window on Monday, and that of a shop girl seen when the same place is re-visited on Tuesday, may combine, in Tuesday night's dream, into a single image of the shop girl wearing the pink dress. But on waking and recalling the dream, the two components of the dream-image, dress and girl, are clearly distinguishable as images of originally separate impressions.

* * *

Concepts.

It will be noticed that, in the foregoing list of definitions, no attempt has been made to delve below that class of thought-process which is styled 'imagery'—a class in regard to which the psychoneural connection suggests itself very readily. Thought processes of a higher order are not yet properly—or even, perhaps, improperly—understood. Our knowledge of these is of the very vaguest description. There appear to be certain generalized ideas called '*Concepts*', such as, for example, those we employ when we think of 'eating', 'playing', 'imagining', or of 'difficulty', 'truth', 'deception', 'difference'; but it is even doubtful whether these may legitimately be herded together under any such single class-name. Compare, for instance, 'eating' with 'difference'. The former idea *may* be no more than the stimulation of the more broadly determinative lines of some extensive pattern in the *plexus* of brain-paths; but the latter may claim a connection with, or share in, every single idea we can formulate.

It is here that the animist is enabled to put up his best fight in defence of the observer's alleged power of intervention. But even here the materialist may claim to have overrun a considerable part of the disputed territory. For the man whose brain has been injured by disease may, apparently, forget what 'eating' is; or may be more than a little hazy regarding the existence of a 'difference' betwixt himself and a grasshopper.

Our present pathway does not take us across this particular battlefield; though we pass within hailing distance of the combatants. From them, however, we may accept the information that concepts are often determinants of the route that attention follows through the associational net. It is hardly possible for the unguided attention to dwell upon any concept without finding itself, a moment later, confronted by a generic or even specific, image clearly related to that main idea.

PART II
THE PUZZLE

CHAPTER VI

In this section, it will be necessary to relate, as briefly as possible, the regrettably dramatic and extremely misleading incidents referred to in the second paragraph of the first chapter (the reader will remember the assurance given therein). It will be noticed that the incidents in question mimicked to perfection many classical examples of alleged 'clairvoyance', 'astral-wandering', and 'messages from the dead or dying'. It will be understood that they are described merely for their illustrative worth, and because they form part of the 'narrative of the actual proceedings involved'. But, from one point of view, these occurrences had a value entirely unique. This was because I was not, as is usually the case in such matters, compelled to take them at second-hand from some 'clairvoyant' or 'medium'—with all the important points left out and a mass of misleading suggestion thrown in. For they happened, one and all, to myself.

*　　*　　*

The first incident provided a very fair example of what might easily have passed for 'clairvoyance'.

It occurred in 1899, when I was staying at an hotel in Sussex. I dreamed, one night, that I was having an argument with one of the waiters as to what was the correct time. I asserted that it was half-past four in the afternoon: he maintained that it was half-past four in the middle of the night. With the apparent illogicality peculiar to all dreams, I concluded that my watch must have stopped, and, on extracting that instrument from my waistcoat pocket, I saw, looking down on it, that this was precisely the case. It had stopped—with the hands at half-past four. With that I awoke.

The dream had been a peculiar one (in ways which have nothing to do with this book), and the net result of it all was that I lit a match to see whether the watch had really stopped. To my surprise it was not, as it usually is, by my bedside. I got out of bed, hunted round, and found it lying on the chest of drawers. Sure enough, it *had* stopped, and the hands stood at half-past four.

The solution seemed perfectly obvious. The watch must have stopped during the previous afternoon. I must have noticed this, forgotten it, and remembered it in my dream. Satisfied on that point, I rewound the instrument, but, not knowing the real time, I left the hands as they were.

On coming downstairs next morning, I made straight for the nearest clock, with the object of setting the watch right. For if, as I supposed, it had stopped during the previous afternoon, and had merely been rewound at some unknown hour of the night, it was likely to be out by several hours.[1]

To my absolute amazement I found that the hands had lost only some two or three minutes—*about the amount of time which had elapsed between my waking from the dream and rewinding the watch.*

This suggested, of course, that the watch had stopped at the actual moment of the dream.[2] The latter was probably

[1] In other words, it was extremely unlikely that I should have dreamed of half-past four at precisely half-past four. A correspondent, Mr. C. G. Newland, points out that I should make this more clear since the question was essentially one of probability.

[2] The improbability of my having dreamed of half-past four *at* half-past four must be multiplied by the improbability of my having been bothered by a stopped watch on the previous afternoon without retaining the faintest recollection of such a fact.

brought about by my missing the accustomed ticking. But—how did I come to see, in that dream, that the hands stood, as they actually did, at half-past four?

If anyone else had told me such a tale I should probably have replied that he had dreamed the whole episode, from beginning to end, including the getting up and rewinding. But that was an answer I could not give to myself. I *knew* that I had been awake when I had risen and looked at the watch lying on the chest of drawers. Yet, what was the alternative? 'Clairvoyance'—seeing across space through darkness and closed eyelids? Even supposing that there existed unknown rays which could effect that sort of penetration, and then produce vision—which I did not believe—the watch had been lying at a level above that of my eyes. What sort of rays could these be which bent round corners?

From Sussex, I went to Sorrento, in Italy. Lying in bed there one morning, I awoke and fell to wondering what the time might be. I lacked energy to look at my watch, which lay outside the mosquito curtains, on a small table within reach, but out of sight when my head was on the pillow. It occurred to me to experiment with the object of ascertaining whether I could again see that watch in the apparently 'clairvoyant' fashion of the earlier experience. Closing my eyes, and concentrating my thoughts upon wondering what the time might be, I fell into one of those semi-dozes in which one is still aware of one's situation. A moment later I found myself looking at the watch. The vision I saw was binocular, upright, poised in space about a foot from my nose, illumined by ordinary daylight, and encircled by a thick, whitish mist which filled the remainder of the field of sight. The hour hand stood at exactly eight o'clock; the minute hand was wavering between the twelve and the one; the second hand was a formless blur. To look more intently would, I felt, wake me completely, so I made up my mind to treat the minute hand as one treats the needle of a prismatic compass, and to divide the arc of its swing. This gave the time as two and a half minutes past eight. That decided, I opened my eyes, reached out under the mosquito curtains, grabbed the watch, pulled it in, and held it up before me. I was wide awake, and—the hands stood at two and a half minutes past eight.

This time there seemed to be no way out. I was driven to the conclusion that I possessed some funny faculty of *seeing*—seeing through obstacles, across space, and round corners.

But I was wrong.

* * *

Then came an incident of an entirely different character.

In January, 1901, I was at Alassio, on the Italian Riviera, having been invalided home from the Boer War. I dreamed, one night, that I was at a place which I took to be Fashoda, a little way up the Nile from Khartoum. The dream was a perfectly ordinary one, and by no means vivid, except in one particular. This was the sudden appearance of three men coming from the South. They were marvellously ragged, dressed in khaki faded to the colour of sackcloth; and their faces under their dusty sun-helmets were burned almost black. They looked, in fact, exactly like soldiers of the column with which I had lately been *trekking* in South Africa, and such I took them to be. I was puzzled as to why they should have travelled all the way from that country to the Sudan, and I questioned them on that point. They assured me, however, that this was precisely what they had done. 'We have come right through from the Cape,' said one. Another added: 'I've had an awful time. I nearly died of yellow fever.'

The remainder of the dream was unimportant.

At that time we were receiving the *Daily Telegraph* regularly from England. On opening this paper at breakfast, the morning after the dream, my eye was caught by the following flaring headlines:

THE CAPE TO CAIRO
'DAILY TELEGRAPH'
EXPEDITION AT KHARTOUM

From our special correspondent.
KHARTOUM, *Thursday* (5 p.m.).
The *Daily Telegraph* expedition has arrived at Khartoum after a magnificent journey, etc., etc.

A note in another part of the paper stated that the expedition was led by M. Lionel Decle.[1] I heard or read subsequently that one of the three white men of the party had died *en route*; not, however, of yellow fever, but of enteric. Whether this was true, or whether there were three white leaders, I do not know.

One or two remarks may be made here.

I had heard, some years previously, that M. Lionel Decle was contemplating some such trans-continental journey; but I did not know that anything had come of the scheme. Certainly I had no idea that the expedition had started.

The expedition arrived at Khartoum the day before the news was published in London, and thus long before I had the dream, as that issue of the paper had to get from London to Alassio, and the dream did not occur till the night before its arrival. This put any 'astral-wandering' business completely out of the question.

I attempted no explanation.

[1] The reader should bear in mind that, in the era of which I write, African exploration was a subject of great interest to everyone. This was the first occasion on which the 'Dark Continent' had been crossed in this direction, and the event was 'news' of the first magnitude.

* * *

The next incident was as dramatic as any lover of the marvellous could desire.

In the spring of 1902 I was encamped with the 6th Mounted Infantry near the ruins of Lindley, in the (then) Orange Free State. We had just come off *trek*, and mails and newspapers arrived but rarely.

There, one night, I had an unusually vivid and rather unpleasant dream.

I seemed to be standing on high ground—the upper slopes of some spur of a hill or mountain. The ground was of a curious white formation. Here and there in this were little fissures, and from these jets of vapour were spouting upward. In my dream I recognized the place as an island of which I had dreamed before—an island which was in imminent peril from a volcano. And, when I saw the vapour spouting from the ground, I gasped: 'It's the island! Good Lord, the whole thing is going to *blow up*!' For I had memories of reading about Krakatoa, where the sea, making its way into the heart of a volcano through a submarine crevice, flashed into steam, and blew the whole mountain to pieces. Forthwith I was seized with a frantic desire to save the four thousand (I knew the number) unsuspecting inhabitants. Obviously there was only one way of doing this, and that was to take them off in ships. There followed a most distressing nightmare, in which I was at a neighbouring island, trying to get the incredulous *French* authorities to despatch vessels of every and any description to remove the inhabitants of the threatened island. I was sent from one official to another; and finally woke myself by my own dream exertions, clinging to the heads of a team of horses drawing the carriage of one 'Monsieur le Maire', who was going out to dine and wanted me to return when his office would be open next day. All through the dream the *number* of the people in danger obsessed my mind. I repeated it to everyone I met, and, at the moment of waking, I was shouting to the 'Maire', 'Listen! Four thousand people will be killed unless——'

I am not certain now when we received our next batch of papers, but, when they did come, the *Daily Telegraph* was amongst them, and, on opening the centre sheet, this is what met my eyes:

VOLCANO DISASTER IN MARTINIQUE

TOWN SWEPT AWAY

AN AVALANCHE OF FLAME

PROBABLE LOSS OF OVER 40,000 LIVES

BRITISH STEAMER BURNT

One of the most terrible disasters in the annals of the world has befallen the once prosperous town of St Pierre, the commercial capital of the French island of Martinique in the West Indies. At eight o'clock on Thursday morning the volcano Mont Pelée which had been quiescent for a century, etc., etc.

But there is no need to go over the story of the worst eruption in modern history.

In another column of the same paper was the following, the headlines being somewhat smaller:

A MOUNTAIN EXPLODES

There followed the report of the schooner *Ocean Traveller*, which had been obliged to leave St Vincent owing to a fall of sand from the volcano there, and had subsequently been unable to reach St Lucia owing to adverse currents opposite the ill-fated St Pierre. The paragraph contained these words:

'When she was about a mile off, the volcano Mont Pelée exploded.'

The narrator subsequently described how the mountain seemed to split open all down the side.

Needless to say, ships were busy for some time after, removing survivors to neighbouring islands.

There is one remark to be made here.

The number of people declared to be killed was not, as I had maintained throughout the dream, 4,000, but 40,000. I was out by a nought. But, when I read the paper, I read, in my haste, that number as 4,000; and, in telling the story subsequently, I always spoke of that printed figure as having been 4,000; and I did not know it was really 40,000 until I copied out the paragraph fifteen years later.

Now, when the next batch of papers arrived, these gave more exact estimates of what the actual loss of life had been; and I discovered that the true figure had nothing in common with the arrangement of fours and noughts I had both dreamed of, and gathered from the first report. So my wonderful 'clairvoyant' vision had been wrong in its most insistent particular! But it was clear that its wrongness was likely to prove a matter just as important as its rightness. For *whence*, in the dream, had I got that idea of 4,000? Clearly it must have come into my mind *because of the newspaper paragraph*. This suggested the extremely unpleasant notion that the whole thing was what doctors call 'Identifying Paramnesia'; that I had never really had any such dream at all; but that, on reading the newspaper report, a false idea had sprung up in my mind to the effect that I had previously dreamed a dream containing all the details given in that paragraph.

Moreover, reflection showed that the Cape to Cairo vision might very well have been of the same character. Indeed, the more I thought of the two episodes the clearer it became that, in each case, the dream had been precisely the sort of thing I might have expected to have experienced *after* reading the printed report—a perfectly ordinary dream based upon the personal experience of reading. How, then, could I be sure that those dreams had not been *false memories* engendered by the act of reading?

But there was the watch business to be taken into account. That, certainly, could not be made to fit in with the new theory, unless I were a great deal madder than I could bring myself to believe.

I was, however, absolutely satisfied that neither in the Cape to Cairo nor in the Mont Pelée dream had there been any 'astral-wandering', or any direct vision across leagues of space, or any 'messages' from the actors in the actual episodes represented. These dreams had been induced, either by the reading of the paragraphs, or else by telepathetic communications from the journalist in the *Daily Telegraph* office who had written those accounts.

CHAPTER VII

To my great relief, the next experience, which occurred some two years later, completely squashed the 'Identifying Paramnesia' theory.

I dreamed that I was standing on a footway of some kind, consisting of transverse planks flanked on my left side by some sort of railing, beyond which was a deep gulf filled with thick fog. Overhead, I had an impression of an awning. But this last was not clearly seen, for the fog partly hid everything except three or four yards of the planking ahead of me with its attendant portion of railing and gulf. Suddenly I noticed, projecting upwards from somewhere far down in the gulf, an immensely long, thin, shadowy thing like a gigantic lath. It reached above the plankway, and was slanted so that it would, had the upper end been visible through the fog, have impinged upon the awning. As I stared at it, it began to wave slowly up and down, brushing the railing. A moment later I realized what the object was. I had seen just such a thing once before in a cinema picture of a fire, in the early days of cinematography. Then, as now, I had undergone the same puzzlement as to what this sort of waving lath might be, until I had realized that it was the long water-jet from a fire-engine hose, as photographed through intervening smoke. Somewhere down in that gulf, then, there must be a fire-engine, and it was playing a stream of water upon the smoke-hidden, railed structure where I stood. As I perceived this, the dream became perfectly abominable. The wooden plankway became crowded with people, dimly visible through the smoke. They were dropping in heaps; and all the air was filled with horrible, choking, gasping ejaculations. Then the smoke, which had grown black and thick, rolled heavily over everything, hiding the entire scene. But a dreadful, suffocated moaning continued—and I was entirely thankful when I awoke.

I was taking no chances with 'Identifying Paramnesia' this time. I carefully recalled every detail of the dream after waking, and not till I had done this did I open the morning papers. There was nothing in these. But the evening editions brought the expected news.

There had been a big fire in a factory somewhere near Paris. I think it was a rubber factory, though I cannot be sure. At any rate it was a factory for some material which gave off vile fumes when burning. A large number of workgirls had been cut off by the flames, and had made their way out on to a *balcony*. There, for the moment, they had been comparatively safe, but the ladders available had been too short to admit of any rescue. While longer ones were being obtained, the fire-engines had directed streams of water on to the balcony to keep that refuge from catching alight. And then there happened a thing which must, I imagine, have been unique in the history of fires. From the broken windows behind the balcony the smoke from the burning rubber or other material came rolling out in such dense volumes that, although the unfortunate girls were standing actually in the open air, every one of them was suffocated before the new ladders could arrive.

This dream left the whole business more puzzling than ever. It seemed that nothing could explain it. For 'clairvoyance' is not an explanation. It is a meaningless expression, a mere admission of inexplicability. And 'telepathy' required an enormous amount of stretching before it could be made to fit the facts.

* * *

Then came a dream which somewhat simplified matters. For it ruled out definitely: insanity, clairvoyance, astral-wandering, spirit-messages, and telepathy. But it left me face to face with something much more staggering than any of these.

In 1904, a few months after the fire dream, I was staying at the Hotel Scholastika, on the borders of the Aachensee, in Austria. I dreamed one night that I was walking down a sort of pathway between two fields, separated from the latter by high iron railings, eight or nine feet high, on each side of the path. My attention was suddenly attracted to a horse in the field on my *left*. It had apparently gone mad, and was tearing about, kicking and plunging in a most frenzied fashion. I cast a hasty glance backwards and forwards along the railings to see if there were any openings by which the animal could get out. Satisfied that there was none, I continued on my way. A few moments later I heard hoofs thundering behind me. Glancing back I saw, to my dismay, that the brute *had* somehow got out after all, and was coming full tilt after me down the pathway. It was a full-fledged nightmare—and I ran like a hare. Ahead of me the path ended at the foot of a flight of wooden steps rising upward. I was striving frantically to reach these when I awoke.

Next day I went fishing with my brother down the little river which runs out of the Aachensee. It was wet-fly work, and I was industriously flogging the water when my brother called out: 'Look at that horse!' Glancing across the river, I saw the scene of my dream. *But, though right in essentials, it was absolutely unlike in minor details.* The two fields with the fenced-off pathway running between them were there. The horse was there, behaving just as it had done in the dream. The wooden steps at the end of the pathway were there (they led up to a bridge crossing the river). But the fences were wooden and small—not more than four or five feet high—and the fields were ordinary small fields, whereas those in the dream had been park-like expanses. Moreover, the horse was a small beast, and not the rampaging great monster of the dream—though its behaviour was equally alarming. Finally, it was in the wrong field, the field which would have been on my *right*, had I been walking, as in the dream, down the path towards the bridge. I began to tell my brother about the dream, but broke off because the beast was behaving so very oddly that I wanted to make sure that it could not escape. As in the dream, I ran my eye critically along the railings. As in the dream, I could see no gap, or even gate, in them anywhere. Satisfied, I said, 'At any rate, *this* horse cannot get out', and recommenced fishing. But my brother interrupted me by calling, 'Look out!' Glancing up again, I saw that there was no dodging fate. The beast *had*, inexplicably, just as in the dream, got out (probably it had jumped the fence), and, just as in the dream, it was thundering down the path towards the wooden steps. It swerved past these and plunged into the river, coming straight towards us. We both picked up stones, ran thirty yards or so back from the bank, and faced about. The end was tame, for, on emerging from the water on our side, the animal merely looked at us, snorted, and galloped off down a road.

Now, it seemed to me that from this incident one thing was abundantly clear. These dreams were not *percepts* (impressions) of distant or future events. They were the usual commonplace dreams composed of distorted *images* of waking experience, built together in the usual half-senseless fashion peculiar to dreams. That is to say, *if they had happened on the nights after the corresponding events*, they would have exhibited nothing in the smallest degree unusual, and would have yielded just as much true, and just as much false, information regarding the waking experiences which had given rise to them as does any ordinary dream—which is very little.

They were the ordinary, appropriate, expectable dreams; but they were occurring on the *wrong nights*.

Even the watch dreams were merely the dreams I ought to have had *after* seeing the watch. In the first of those incidents I had, when awake, seen the watch lying *face upwards* on the chest of drawers, with the hands stopped; and the corresponding dream image had been of a stopped watch, face upwards. In the second instance I had *held the watch up facing me about a foot from my nose, while lying with my head on my pillow*; and the reader will remember that the corresponding dozing image had been of a watch in precisely that position. The white mist had been, of course, the image of the mosquito curtains, out of focus, as these were, when I looked at the real watch.

No, there was nothing unusual in any of these dreams as dreams. They were merely *displaced in Time*.

That, of course, was staggering enough. But I felt, nevertheless, that it had been a great advance to resolve all these varied phenomena into one single class of incident—a simple, if mysterious, transposition of dates.

But in all this speculation I was still a long way from the truth.

The two remaining incidents I propose to relate in this section contained nothing to alter my half-formed opinion that temporal aberration constituted the whole of the mystery involved. But, had I not made this semi-discovery, I should certainly have regarded the following incident as a message from the 'spirit-world' or a 'phantasm of the dying'.

* * *

In 1912 I spent a good deal of time at Salisbury Plain, experimenting with one of my stable aeroplanes. A military aeroplane competition was in progress, and most of the officers of the then tiny Royal Flying Corps were there. One of these I had not met before, nor did I see very much of him; in fact, I do not think I spoke to him more than twice. Since these records are not evidence, or intended to be regarded as such, it will suffice if I refer to him as Lieutenant B. The other officers were all old friends of mine. Shortly after the conclusion of the competition the annual army manœuvres began, and, having nothing to do with these, I went to Paris to inspect another machine which was being built there to my design.

One morning while in that city I dreamed that I was standing in a very large meadow, situated in a landscape which I did not recognize. In this meadow a monoplane landed, crashing rather badly some fifty yards away. Immediately afterwards I saw B. coming to me from the direction of the wreck. I asked if much damage had been done. He replied, 'Oh no, not much,' and then added, 'It's all that beastly engine; but I've got the hang of it now.' The dream was a longish one, all about aeroplane accidents (a common form of nightmare with me, even to this day), and B.'s smash was by no means the worst thing I saw. I awoke to find the servant by my bedside with the morning tea, from which fact I was subsequently able to fix the hour of the dream as close on 8 a.m.

B. was killed between 7 and 8 that morning, falling into a meadow near Oxford. But I did not read of the accident till about two days and a night later.

But now, note the following points:

1. Engine failure had nothing whatever to do with the accident, nor could B. for one moment have ever

thought that it had. For the monoplane was planing down—with the engine partly or entirely stopped—at the time; and the accident was due to the uncoupling of a quick-release gadget in one of the main 'lift' wires, and the consequent breaking upward of one wing. Of course, the planing down may have been compulsory, and due to engine failure; but there could have been no doubt in B.'s mind that his wing had broken.

On the other hand, B. had made to my sister, while we were at the Plain, a remark about the engine almost exactly like that I heard him make in the dream, and it is more than likely that she had repeated it to me. She would naturally have done so.

2. B. was merely a passenger in the machine. It was being piloted by another man, a stranger to me, who was also killed. There was nothing of this in the dream.

But when I read the paragraph about the catastrophe, it was B.'s name alone which caught and held my attention; and I did not know of the death of the other man until I looked up the record of the accident several years later.

3. The paragraph did not state the cause of the accident, and so left me with nothing to go upon but (possibly) B.'s past remark about the engine.

4. The coincidence in time was not really remarkable. Dreams of aeroplane accidents were, as I have said, very frequent with me in those days, and between seven and eight, when the noise of motor traffic in the streets begins to penetrate to one's consciousness, has always been my hour for this particular class of nightmare.

So I concluded that here, again, the dream was associated with the personal experience of reading the paragraph.

* * *

In the last incident of this series, the chronological aberration was far more considerable.

The dream occurred in the autumn of 1913. The scene I saw was a high railway embankment. I knew in that dream—knew without questioning, as anyone acquainted with the locality would have known—that the place was *just north of the Firth of Forth Bridge*, in Scotland. The terrain below the embankment was open grassland, with people walking in small groups thereon. The scene came and went several times, but the last time I saw that a train going north had just fallen over the embankment. I saw several carriages lying towards the bottom of the slope, and I saw large blocks of stone rolling and sliding down. Realizing that this was probably one of those odd dreams of mine, I tried to ascertain if I could 'get' the date of the real occurrence. All I could gather was that this date was somewhere in the following spring. My own recollection is that I pitched finally upon the middle of April, but my sister thinks I mentioned March when I told her the dream next morning. We agreed, jokingly, that we must warn our friends against travelling north in Scotland at any time in the succeeding spring.

On April the 14th of that spring the 'Flying Scotsman', one of the most famous mail trains of the period, jumped the parapet near Burntisland Station, about fifteen miles north of the Forth Bridge, and fell on to the golf links twenty feet below.

* * *

The above-described incidents have been selected from a group of about twenty, simply because they were closely studied and carefully memorized at the time of their occurrence. Most of the others were merely noted, so to say, *en passant*, and are now almost completely forgotten.

Curiously, I can remember no dreams of the coming Great War—except one. That one related to the bombardment of Lowestoft by the German fleet. I recognized the place as Lowestoft, but had no idea of the nationality of the bombarding vessels.

PART III

THE EXPERIMENT

CHAPTER VIII

No-one, I imagine, can derive any considerable pleasure from the supposition that he is a freak; and, personally, I would almost sooner have discovered myself to be a 'medium'. There might have been a chance of company there. Unfortunately it was abundantly clear that there was no 'mediumship' in this matter, no 'sensitiveness', no 'clairvoyance'. I was suffering, seemingly, from some extraordinary fault in my relation to reality, something so uniquely wrong that it compelled me to perceive, at rare intervals, large blocks of otherwise perfectly normal personal experience displaced from their proper positions in Time. That such things could occur at all was a most interesting piece of knowledge. But, unfortunately, in the circumstances it could be knowledge to only one person—myself.

There was, however, a very remote possibility that, by employing this piece of curiously acquired knowledge as a guide, I might be able to discover some hitherto overlooked peculiarity in the structure of Time; and to that task I applied myself.

Progress here was definite, but it was terribly slow. There was no help to be found in the conception of Time as a fourth dimension. For Time has always been treated by men of science as if it were a fourth dimension. What had to be shown was the possibility of *displacement* in that dimension. Nor did I gather much comfort from Bergson;

for to tell a man who is confronted with parts of Time clearly transposed that Time has no parts is distinctly futile. I cared not a whit whether Time were 'a form of thought', or an aspect of reality, or (this was later) compoundable with Space. What I wanted to know was: How it got *mixed*?

For 'mixed' was the right word. Between the dream and the corresponding waking experience came the memory of the dream, while the memory of the waking experience followed them all!

However, the coming of the first world war put a temporary stop to further investigation; and it was not until 1917 that any new developments occurred.

In January of that year I was in Guy's Hospital, recovering from an operation. There, one morning, when reading a book, I came upon a reference to one of those 'combination' locks which are released by the twisting of rings embossed with letters of the alphabet. As I read this, something seemed, for one fleeting instant, to be stirring, so to say, in my memory; but, whatever it was, it immediately subsided. I paused for a second, but nothing further developed, so I returned to my book. Then, luckily, I changed my mind, tossed the volume aside, and set myself determinedly to worry out exactly what it was that I had momentarily associated with the sentence read. In a little while it came back. I had dreamed, during the previous night, of precisely such a combination lock.

The chances of coincidence, where two such vague, commonplace events were concerned, needed no pointing out. But I could not remember having seen, heard, or thought of such a lock for a year or more. And, knowing from past happenings that my dreams did, sometimes, contain images of future experience, it seemed to me that the appearance of the lock image in the previous night's dream might have been another instance of my particular abnormality. Such a supposition might prove, at any rate, worth considering.

A few days later the great Silvertown explosion occurred, shaking the whole building, breaking windows, and causing the nurses to extinguish the lights, on the supposition that Zeppelins were overhead. Such an experience was calculated to make one dream; and dream I did, but, as usual, on the wrong night—the night before the associated event. After the disaster I told a fellow-convalescent of this experience. He interrupted me, saying, 'Wait!' and then: 'Curious, that. Now that I come to think of it, I also dreamed of an explosion last night.'

He could no longer, by then, recall any of the details of his dream, and, since big bangs of all sorts were fairly common during the war, coincidence might well have been responsible for the facts. But—supposing this were not the case, and that the dream had been in the same class as mine? What followed?

There were thus two new suppositions to be examined. Viewed separately, each of these appeared wild in the extreme; but considered together they were sufficiently suggestive to justify a little closer attention.

The validity of the first of these would mean that my dream pre-images were connected, not only with highly exciting and dramatic events, but also with the veriest trivialities, such as this little matter of reading about a combination lock. Exactly, in fact, as dream images of past events are connected just as often with unimportant happenings as with experiences more striking. Again, it had been by the merest accident of fortune that I had set myself to recall that dream; and had I not done so I should never have been aware of the incident. According to this, then, I might, for all I could tell, have had these dreams with considerable frequency, and have either forgotten them at once, or else have *failed to notice their connection* with the subsequent related events.

But, if the supposition about my friend's dream were correct, *this failure to observe a connection was precisely what had happened in his case.* He had not completely forgotten the dream, but the occurrence of the actual explosion had not served to recall it.

I had got no further than this in my speculations when the friend in question came up in a state of some excitement. 'You remember what we were saying about dreams?' he asked. 'Well, I have been talking to So-and-So' (one of the hospital surgeons), 'and he told me of a curious thing which had happened to him the other night. He had just got into bed and gone to sleep when he dreamed that he was aroused and compelled to go out to attend to a fractured leg. Almost immediately after his dream he *was* aroused, owing to the arrival of an urgent message which necessitated his going out to attend to just such a case. And in telling me the story he pointed out that he had not had to deal with a fractured leg for over six weeks.'

So here, possibly, was a third incident, involving a third person. What, I wondered, would become of the record of that event? The surgeon would tell it to a few friends, who would attribute the whole thing to coincidence (it *might* have been that), and in course of time he would forget all about it himself. But——

And then, what about that curious feeling which almost everyone has now and then experienced—that sudden, fleeting, disturbing conviction that something which is happening at that moment *has happened before*?

What about those occasions when, receiving an unexpected letter from a friend who writes rarely, one recollects having dreamed of him during the previous night?

What about all those dreams which, after having been completely forgotten, are suddenly, for no apparent reason, recalled later in the day? *What is the association which recalls them?*

What about those puzzling dreams from which one is awakened by a noise or other sensory event—dreams in which the noise in question appears as the final dream incident? Why is it that this closing incident *is always logically led up to by the earlier part of the dream*?

What, finally, of all those cases, collected and tabulated by the Society for Psychical Research, where a dream of a friend's death has been followed by the receipt, next day, of the confirmatory news? Those dreams were, clearly, not 'spirit messages', but instances of *my* 'effect'—simple dreams associated merely with the coming personal experience of *reading the news*.

I had done nothing but suppose, in hopelessly unscientific fashion, for a week or more, and it seemed to me that I might as well complete my sinning. So I took a final wild leap to the wildest supposition of all.

Was it possible that these phenomena were not abnormal, but *normal*?

That dreams—dreams in general, all dreams, everybody's dreams—were composed of *images of past experience and images of future experience blended together in approximately equal proportions*?[1]

[1] The present reader, doubtless, has grasped the fact that this section of the book is purely historical. On that day in 1917, I was trying to formulate for myself some statement of the possible facts which would serve as a basis for an experimental investigation, and I am describing here the sequence of the ideas which flashed through my mind. The suspicion of an equal distribution of precognitive and retrospective elements came first, and was followed immediately by the more rational theory set forth in the next paragraph, a theory which made the distribution depend upon associational factors which would vary with each individual and in each dream. As will be seen in the next three pages, even this first approximation to the truth was set aside as 'obviously incomplete'. The theory finally accepted was not developed until 1926, and is described in the last section of the book.

That the universe was, after all, really stretched out in Time, and that the lop-sided view we had of it—a view with the 'future' part unaccountably missing, cut off from the growing 'past' part by a travelling 'present moment'—was due to a purely mentally imposed barrier which existed only when we were awake? So that, in reality, the associational network stretched, not merely this way and that way in Space, but also backwards and forwards in Time; and the dreamer's attention, following in natural, unhindered fashion the easiest pathway among the ramifications, would be continually crossing and recrossing that properly non-existent equator which we, waking, ruled quite arbitrarily athwart the whole.

The foregoing supposition was not, be it noted, perceived as a possible *explanation*. The mixture in the order of actual experience—*viz.*, dream, memory of dream, corresponding waking impression, and memory thereof—would still have to be accounted for. But it would put the problem on an entirely different footing. There would be no longer any question as to why a man should be able to observe his own future mental states; that would be normal and habitual. On the contrary, the initial puzzle would be: What was the *barrier* which, in certain circumstances, debarred him from that proper and comprehensive view?

All this was seen in, so to say, a single flash of thought, almost too rapid for analysis.

It was rejected with even greater swiftness. For it was absolutely inconceivable that a thing of this sort, if true, could have managed to escape, through all these centuries, universal perception and recognition.

CHAPTER IX

A little later on, however, I saw that this abrupt recoil had been illogical. For the whole supposition had been based, of course, upon the earlier hypothesis that any general recollection of these images was rendered difficult by the species of inhibition which had prevented my friend from associating his waking experience of the explosion with his previous dream. No memory is ever aroused unless there is some associated idea which revives it, and if that association misses fire, there can be no recall.

Dreams, moreover, are mostly about trivial things—things which happen every day of one's life. Such a dream, even if it were, in actual fact, related to tomorrow's event, would naturally be attributed to yesterday's similar incident. Then, again, nine-tenths of all dreams are completely forgotten within five seconds of waking, and the few which survive rarely outlast the operation of shaving. Even a dream which has been recalled and mentally noted is generally forgotten by the afternoon. Add to this the before-mentioned partial mental ban upon the requisite association; add to that an unconscious, matter-of-fact assumption of impossibility; and it becomes quite probable that it would be only a very few of the more striking, more detailed, and (possibly) more emotional incidents which would ever be noticed at all. These, moreover, would be attributed to telepathy or to 'spirit messages', or even to anything which, though insane in other respects, could, at least, be expressed in the conventional terms of a single absolute, one-dimensional Time.

It was true, of course, that the theory of normality

would take a lot of threshing out. The statement made in the last chapter was, obviously, incomplete; and the full description of the process involved might never be forthcoming. But the alternative was the hypothesis of abnormality; and that meant, not merely abnormality in the sense of excess of, or deficiency in, some common quality of mind, but abnormality in a sense which was itself senseless. It is difficult really to believe in the utterly meaningless.

Finally (and this was what attracted me most), the supposition of normality—of something inherent, not in this or that individual, but in Time itself—would mean, if correct, that, if only one could devise an experiment which would overcome the *two initial difficulties of remembering and associating*,[1] the thing might prove to be directly observable by a very large number of people, including the present reader.

The arrangement of that experiment was, clearly, the first step. Explanation could come (and, as will be seen, did come) later.

[1] The difficulty of remembering is easily overcome; but the difficulty of associating proves in some cases insurmountable. It is always hard to discover in the average dream any incident which is clearly related to a *chronologically definite* past waking event, and some people's dreams are far too complex to allow such connections to be traced. It is obvious that persons thus handicapped would find it equally impossible to discover in their dreams any clear suggestion of precognition.

CHAPTER X

[*Note to Third Edition.*

The instructions given in this chapter are, one and all, of extreme importance. Indeed, it may safely be said that, unless the reader follows them in every detail, he will be reducing his chances of getting results almost to vanishing point. He should bear in mind that while millions of persons remember some of their dreams, and hundreds have written them down, yet not one in a thousand through all the past centuries seems to have *noticed* that he dreams of the future. Obviously, then, it will be useless for him to experiment upon any old-fashioned lines—some entirely novel technique is required. That technique is explained here. But experience since the publication of the book shows that its importance was not sufficiently emphasized. I have added, therefore, to this chapter, as previously written, several pages of more detailed explanations.]

The reader will have guessed that the experiment referred to in the last chapter was tried, and that it proved successful; because otherwise, manifestly, this book would never have been written.

It was not, however, until the following winter that I could bring myself to take the normality hypothesis seriously enough to put it to the test. Then, with many misgivings, and practically no hope of success, I began the first essential experiment, upon myself. I knew, of course, that I had these dreams occasionally; but only at intervals of sometimes a year or more. According to the new theory, however, I should be having similar dreams throughout all these intervals, unknown to myself.

As a rule, on nine mornings out of ten, I have no recollection of having dreamed at all. That, however, did not greatly trouble me. Many people, I knew, were genuinely convinced that they never dreamed; but, from experiments I had made, I was satisfied that 'dreamless sleep' is an illusion of memory. What happens is that one forgets the dreams at the very instant of waking. I myself have remembered, some days later, a dream which had occurred when I was under an anæsthetic, although, during the intervening interval, I had believed myself to have been, at the time, in a state of complete unconsciousness.

My starting-point, then, was a belief in the possibility of recalling a fraction of the lost dreams of these apparently blank nights of mine. Now, according to the new hypothesis, that fraction could contain images of both past and future events. *It was probable that the majority of such images would not be distinct and separate, but, on the contrary, so blended and intermingled that the components would not be distinguishable as belonging to any special waking event.* But just as one can, occasionally, clearly identify one part of such a blend of images as relating to a particular past event (*vide* definition of 'Integration' in Part I), so should one be able, on occasion, to identify an element in the blend as pertaining to a particular future occurrence. The point was (and this is an important point) that one must not expect ever to come upon a complete idea or scene which related *wholly* to the future. As an example of what I mean, the reader may turn back to the dream of a horse, narrated in Part II. There, the greater part of the dream related to the future; but the general *appearance* of the horse, and that of the fields and railings, were, to the best of my belief, details collected from past experience.

The dream, if recalled, would preferably be written

down, so as to make the remainder of the experiment a matter of comparison between two hard, material facts—the record and the waking event. And, to facilitate subsequent analysis of the dream-images, these would best be described with as much detail as possible. A short record, full of detail, would be of more value than a long one drafted in vaguer terms.

But there was an even more cogent reason why amplitude of detail would be essential. A long dream contains a great many images, and a long day a great many impressions. By the ordinary laws of chance some of these would be bound to fit, if the experiment were sufficiently extended. Hence corroborative detail would have to be the crucial test. For example, the dream of a pile of coins on a book, followed next day by the observation of a pile of coins in such a position, would be of the class of coincidence which would be bound to occur in any case. What would be required would be something more in the nature of a pile of *sixpences upsetting* off a *red* book, followed by such a waking experience. (The rest of the scene of such a dream—the table and the room and the cause of the mishap—would probably be entirely different; but that would not matter.) The point was that nothing should be accepted as relating clearly to the future which did contain the elements of what a racing man would call a 'double event'.

The next thing to be considered was the necessity of a time limit. Obviously, even a dream of a pile of sixpences upsetting off a red book would be likely to be matched by a similar waking experience, if one allowed oneself the whole of one's life in which to look for the matching. A bank clerk might even find fulfilment in a fortnight. I decided that two days should be the accepted limit; *but that this might be extended in ratio to the oddity and unusualness of the incident.* That would be a matter for judgement. My dream of the bombardment of Lowestoft, for instance, occurred a year or so before the event; and I have had one clear case—to be described later—of a dream-image relating beyond all possibility of doubt to an event which happened some twenty years later.

Since, then, the possibility of satisfactory identification would depend mainly upon unusualness in the incident, the worst time to choose for the experiment would be the period when one was leading a dull life with each day exactly like the last. But in such circumstances a visit to a theatre or to a cinema might well prove a useful auxiliary to the experiment. (That, I may say now, is an invaluable tip.) Also, one might expect to get dreams of novels one was going to read. (I may add here that one does, as a matter of fact, get some of one's best results that way.) *But, speaking generally, it would be best to select nights preceding a journey or some other expected break in the monotony of circumstances.*

Another factor would be evidently the *number* of the results achieved. Satisfaction might be obtained either from the previous dreaming of a single, very unusual incident; *or equally well from the previous dreaming of several fairly unusual events, any one of which results, had it been the only one, might justly have been attributed to rather exceptional coincidence.*[1] So it was decided that all results of the singly decisive kind should be marked with a +; and that results which, though nearly decisive, required the backing of other similar results, should be marked with a sort of hot-cross-bun, thus: ⊕[2]

[1] It is extraordinary how many people overlook this. If the chances of a given coincidence occurring within a certain period are one in a thousand, the chances of a second equally improbable coincidence occurring in the same period are $\frac{1}{1000} \times \frac{1}{1000}$, or one in a million.

[2] In my own records I marked with a plain circle any dream which appeared to be related to some chronologically definite incident of the past. And I sought for similar evidence of retrospection in the experiments of my assistants.

The foregoing describes the conditions I laid down for the test, and also the nature of the difficulties I was prepared to encounter. And encounter these I did, in abundance. But there were two which I did not foresee.

The dreaming mind is a master-hand at tacking false interpretations on to everything it perceives. For this reason, the record of the dream should describe as separate facts, (a) *the actual appearance of what is seen, and* (b) *the interpretation given to that appearance.*

For example: during one of the days of the test I happened to be blowing a wood fire with a pair of bellows, and, in so doing, I brought the nozzle of the instrument into contact with the red-hot surface (facing me) of a large log. I do not know whether the reader has ever done this; but the effect is most startling, not to say alarming. A dense shower of very brilliant sparks—a regular Crystal Palace firework display—leaps from the fire straight into your face and goes streaming past your ears, causing you to jump back for fear of being blinded. But there appears to be no heat in these sparks—at any rate, no holes are burned in your clothes. The experience is a most striking and unusual one; and, as it happened, precisely such a shower of sparks had flown past my ears in a dream during the previous night. But I had omitted to record the immediate dream-impression, which was simply that of a shower of little sparks, and had written down, instead, the explanation I had *subsequently* attached to that shower—*viz.*, that a crowd of persons who happened to be present in the dream had been throwing cigarette ends. Both aspects of the dream-incident should have been recorded: first, the image seen, and then the interpretation attached thereto. This should be done throughout all the records.

The second difficulty is one which demands careful attention. For it was here, at last, that I found the thing

I had been looking for—the reason why this curious feature in the character of temporal experience has managed, through all these centuries, to escape universal observation.

The waking mind refuses point-blank to accept the association between the dream and the subsequent event. For it, this association is the *wrong way round*, and no sooner does it make itself perceived than it is instantly rejected. The intellectual revolt is automatic and extremely powerful. Even when confronted with the indisputable evidence of the written record, one jumps at any excuse to avoid recognition. One excuse which is nearly always seized is the dissimilarity of the adjacent parts of the scene, or the fact that there are parts in the 'integration' which do *not* fit the incident; matters which do not, of course, in the least affect the fact that there are parts of the scene or integration which *do* fit with the required degree of exactitude.

The result is that, on reading over the record at the end of the succeeding day (or two days), *one is apt to read straight on through the very thing one is looking for, without even noticing its connection with the waking incident.* The reading should therefore be done slowly, with frequent pauses for consideration and for comparison with the day's events. In the cases of nearly all the results I am going to relate, the connection was, at first, only half glimpsed, *was then immediately rejected*, and was finally accepted only on account of the accumulating weight of the previously unnoticed points of corroborative detail.

The simplest way to avoid this initial failure to notice is to pretend to yourself that the records you are about to read are those of dreams which you are going to have *during the coming night; and then to look for events in the past day which might legitimately be regarded as the causes of those dreams. This is not unfair. It is only a device to enable you to notice; not a device to assist you to judge. That you do later, concerning yourself then solely with the corroborative details, and giving no thought to the Time order.*

* * *

The dodge for recalling the forgotten dreams is quite simple. A notebook and pencil is kept under the pillow, and, *immediately* on waking, before you even open your eyes, you set yourself to remember the rapidly vanishing dream. As a rule, a single incident is all that you can recall, and this appears so dim and small and isolated that you doubt the value of noting it down. Do not, however, attempt to remember anything more, but *fix your attention on that single incident, and try to remember its details.* Like a flash, a large section of the dream in which that incident occurred comes back. What is more important, however, is that, with that section, there usually comes into view an isolated incident from a previous dream. Get hold of as many of these isolated incidents as you can, neglecting temporarily the rest of the dreams of which they formed part. Then jot down these incidents in your notebook as shortly as possible; a word or two for each should suffice.

Now take incident number one. Concentrate upon it until you have recovered part of the dream story associated therewith, and write down the briefest possible outline of that story. Do the same in turn with the other incidents you have noted. Finally, take the abbreviated record thus made and write it out in full. Note *details*, as many as possible. *Be specially careful to do this wherever the incident is one which, if it were to happen in real life, would seem unusual; for it is in connection with events of this kind that your evidence is most likely to be obtained.*

Until you have completed your record, do not allow yourself to think of anything else.

Do not attempt merely to remember. Write the dream down. Waking in the middle of the night, I have several times carefully memorized my preceding dreams. But, no matter how certain I have been that those memories were firmly fixed, I have never found one shred of them remaining in the morning. Even dreams which I have memorized just before getting up, and rememorized while dressing, have nearly always vanished by the end of breakfast.

It will be impossible, of course, for you to write down *all* the detail. To describe the appearance of a single dream-character completely would keep you busy for ten minutes. But write down the general detail, and *all uncommon detail.* Memorize the remainder by reading through your final record and attentively revisualizing each picture described therein; so that, should one of these unwritten details subsequently prove important, you can be satisfied that you are not then recalling it for the first time.

If, on waking, you are convinced that you have not dreamed at all, and cannot recall a single detail, stop trying to recollect the dream, and concentrate, instead, on remembering what you were *thinking* when you first awoke. On recalling that thought, you will find that it was consequent on a dream, and this dream will immediately begin to return.

Read your records over from their beginning at the end of each day of the experiment.

The sort of thing you may expect to find will be described in Chapter XI (*b*).

* * *

[*Note to Third Edition.*

I append here the more detailed explanation referred to in the introductory note to this chapter.

In the experiments to be narrated it was found, to begin with, that the great bulk of the dreams exhibited *no* resemblances to any chronologically definite incident of waking life—past or future. This was entirely contrary to the popular supposition. The very small residue consisted of resemblances to incidents which were distinctively past only or distinctively future only; but these resemblances were mostly too slight to be evidential. However, a closer study of some of these apparently trivial coincidences would bring to light *previously unnoticed corroborative details* which rendered the dream evidential of retrospection or of precognition. Thus, though all dreams were clearly related to waking life as a whole, it would be extremely difficult for anyone to prove, by actual experiment, whether they related to the past or the future or both. Evidence, in either direction, was about equally rare.

But that evidence was not equally difficult to notice. Attention would be arrested at once by the most trivial resemblances to the past, while passing over similar resemblances to the future with scarcely a pause. And the reason was obvious. In the case of a resemblance to the past, a *causal connection* is presupposed; so that the feeble character of the resemblance is ignored, and the dream record is regarded as meriting further examination. But in the case of a resemblance to the future, the degree of resemblance is the *only* evidence of a causal connection hostile to common sense, so that the judgement demands a far higher degree of resemblance before it will regard the incident as *worth considering*. Now, this would not matter, if the resemblances of dreams to waking events leapt to the eye all complete, with every detail in full view and readily estimable at its proper value. *But that, practically, never happens*. The resemblance dawns on one piecemeal; one very trivial similarity is noticed first, and, *if* the judgement is arrested by this, the dream is re-read and the corroborative details come slowly and singly to light.

And, for the reasons already given, this all-important, first, feeble resemblance is promptly—almost unconsciously—dismissed as too far-fetched to merit further consideration, *if it relates to the future.*

This psychological trap is essentially a trap for the expert, the man who realizes how very feeble that first trivial resemblance is. The neophyte is apt to escape it by giving the resemblance a greater value than it possesses.

In short, to notice that a resemblance between a waking event and a past dream is worth following up, is like trying to read a book while looking out for words which might mean something spelled backwards. The mind cannot keep that up for long. One must divide the task—see, first, how the book reads in the ordinary way, and then—*hold it up to a mirror.* Consequently, in the instructions to experimenters given in this book, it is laid down that the subject will have little chance of *noticing* the results he has actually obtained, unless he tries this 'mirror' device, *i.e.*, pretends to himself that the dreams which he has recorded are those which he is going to have on the following night, and then examines the day's events for anything which might be regarded as the cause of those dreams. He is carefully warned, of course, that this is not a device to enable him to judge the value of the evidence: it is a trick to enable him to notice that there is any evidence to be judged.

In a recent article, Sir Herbert Barker referred to this as the most important of all the rules in experiments of this description, and I entirely agree with his finding.

I must re-emphasize here the importance of the advice regarding the choice of nights upon which to experiment, *viz.*, that these should precede some coming break in the monotony of your everyday life. In the experiments to be narrated, Miss B., Miss C., Major B. and myself were holiday-making in entirely new scenes, and obtained dreams resembling events which were distinctively past or future within the allotted period. Mrs L., on the other hand, was living her normal life in her home. Her records were longer than those of all the other experimenters added together, yet she had only one dream resembling a chronologically definite incident of the future, and only one resembling a similarly definite incident of the past.]

CHAPTER XI (*a*)

[*Added to Third Edition.*]

It may simplify matters for the reader if I explain in more detail what it was that, at this stage, I was trying to ascertain.

The picture of the universe which, towards the end of last century, was accepted by almost every class of thinker, was painted in terms of the conventional 'elementary indefinables', 'Space' and 'Time'. Physics had added a third term, 'Matter', and was suffering considerable perplexity as to how, with these three alone, it was going to absorb 'Radiation'. Biology had elected, rather meekly, to consider itself a branch of this particular physics. *Sense data* were regarded as improprieties. The actual result was very much like the patchwork which an ingenious person might construct after mixing together the pieces of several 'jig-saw' puzzles and dropping half upon the floor. It was extraordinarily good in parts, but the parts did not fit.

We know now that the discordances in, at any rate, the physical section were due to our imperfect manner of employing the indefinables of Space and Time. But the hall-mark of that period was an impatience incapable of considering the possibility of errors of so fundamental a character. And it must be remembered that Planck's voice had only just been raised, and that Einstein had not yet spoken.

Supposing, now, that a man of that time had experienced a series of dreams similar to those narrated in the earlier part of this book; he would have discovered something flatly opposed to the conventional view of Time. And that view was sacrosanct: the whole supposedly unassailable structure of physics bore witness to its accuracy. In these circumstances, our hypothetical dreamer would have been compelled to take refuge in Mysticism. He would have had to accept the existence of two disconnected worlds, the one rational, the other irrational.

But by 1917 the situation had changed entirely. The one thing that I did *not* need to worry about was the classical theory of Time. That, already, was in the melting-pot. Modern science had put it there—and was wondering what to do next.

Now, the probabilities that the whole series of dreams already described had been due to pure coincidence were so excessively minute that, *taking into account the partial collapse of the classical theory of Time,* I was bound to postulate precognition as a working hypothesis. Then, as a disciple of science, I must assume, pending absolute proof to the contrary, that precognition was scientifically possible, *i.e.*, that the nature of Time allowed the observer a four-dimensional outlook on the universe. That was eminently reasonable; for, if modern science insisted upon the reality of its four-dimensional 'space-time' (*vide* later chapters), it could not dispute that observers in that world must be similarly four-dimensional. But that would involve that everyone possessed precognitive faculties. Unfortunately, it did not follow that he would employ them. It was possible to enumerate many personal factors which might make retrospection more attractive to the dreamer. And here was the difficulty. To establish my case I should have to overcome the objections of those who would urge, *as a matter of common knowledge,* that dreams which offered a resemblance to the future strong enough to arouse a suspicion of precognition were *not* vouchsafed to the multitude, but were, on the contrary, the prerogative of a few rare individuals.

I should like the reader to be quite clear about the nature of this obstacle. In science, one uses the word 'effect' when one wishes to consider a phenomenon apart from any presumptions as to its possible cause. The strong 'effects' to which I have just referred might or might not be due to coincidence, but that was not the difficulty. The objection which I should have to meet was not that the strong 'effects' were inconclusive evidence of precognition; it was the far more formidable assertion that only an abnormal few could observe any such effects at all!

Now, if I were right, and there remained a still unsuspected logical fallacy in our notions of Time, that fallacy would prove, of course, self-evident—once it was discovered. Moreover, the discovery could hardly fail to affect every branch of science and to reap its quota of confirmation from each. The inexact evidence of dreams could provide no part of the essential basis of a serious scientific theory, and to attempt to make it such would be the worst possible policy. But I could not *ignore* that evidence. My opponents would be able to point out that the existence of universal faculties for dream precognition was a necessary *corollary* of my proposition, and they would demand to know why it was that not one person in a thousand utilized these supposed opportunities. 'The evidence of dreams', they would say, 'is extremely relevant to your theory. And that evidence is flatly against you.'

In these circumstances, it seemed inadvisable to expend further energy upon the extremely difficult Time problem until I had satisfied myself that the striking effects in question were far more widely distributed *among individuals* than the popular view supposed.

Closely allied to that popular view were the opinions of those who believed that precognition was possible, but held that it must involve the employment of an extra, 'supernormal' faculty. This notion was cherished by mystics of every class; and these were likely to raise considerable outcry at the suggestion that their stronghold, sacred for centuries, was open to invasion by mundane science. Unfortunately, they received strong support from some of the people who had devoted most time to the investigations of previsional phenomena, *viz.*, members of the various groups engaged in what is called 'Psychical Research'.

It is interesting to note the curious consequences of this creed. The supernormalist sets himself a certain standard (varying according to taste) beyond which he would rule out coincidence as too improbable. Suppose that one of Jones's dreams attains to this standard—he is credited with having exercised his 'supernormal' faculty. Suppose that Smith has a dream which is very nearly, but not quite, up to that standard—Smith's dream is adjudged to be due to the exercise of the normal dream faculties. *But now suppose* that Jones has a dream similar to Smith's. There is nothing for it but to say that, on this occasion, Jones neglected to exercise the superior faculty. So the change over from one faculty to the other occurs when there is a shade of difference in the odds against coincidence!

Nonsense! did you say? Of course it is. Then how do the supernormalists get over the difficulty? I do not know. They do not appear to notice it. When one of them settles down to the practical work of studying such dreams as he may have collected from the community at large, he grades all resemblances to the future as good, fair, moderate or indifferent. The indifferent he judges to be due to the usual, normal faculty; the good (those upon which he has based his belief) he regards as probably produced by the other and supernormal faculty; the intermediates he sets aside as doubtful. But he forgets entirely that the existence of these intermediate effects compels him to consider that *the supposed change of faculty occurs at some particular point in the scale*—a point between two dreams of nearly similar evidential value.

In short, the only consistent supernormalists—the only ones who avoid the above absurdity—are those who adhere to the popular view that there are *no* intermediates in the scale, that the effects upon which they have based their belief are in a class by themselves, isolated by a wide gap from such inferior effects as can be observed. These persons usually accept the further popular supposition that the effects which are worth counting pertain to rare and specially favoured individuals.

I trust I have made it clear that the object of the projected experiments was to see whether the evidence of dreams in general was really for or against the theory that the faculty of precognition, if it existed, was a normal characteristic of man's general relation to Time. I hoped, in other words, to be able to turn the tables upon objectors of the classes cited above, and to show that effects suggesting precognition were observed by far too many people to allow us to entertain the supposition that these persons differed from their fellows in some supernormal fashion.

* * *

It is obvious that all such effects as might be discovered would have a certain value as evidence of the *fact* of precognition—an aspect to be distinguished from that of their evidence as to the *distribution* of a precognition assumed to exist. It was with the latter aspect that I was concerned; but the former may be of interest to the reader, and he may, indeed, consider that I ought to make some statement concerning my attitude towards such evidence. Very well, I will do my best.

In the first place, of course, we have to recognize that there are *no* limits to the possibilities of coincidence; consequently, evidence of precognition is of a purely statistical character—a matter of balancing probabilities. We are not dealing with an exact science, but with a method which approximates steadily towards exact science as the probabilities grow higher.

Now, the chances against a series of effects being coincidences depend upon two factors, *viz.*,

(1) The oddity of the individual effects.

(2) The frequency of their occurrence.

The *dilettante*, as a rule, overlooks this second factor entirely. Yet the evidence of seven dreams in a given period, with the probability of coincidence in each case as high as one in ten, is actually ten times as strong as the evidence of a single dream with chances of coincidence as low as one in a million.

Let us consider the first of the above factors. If the supernormalists are right—if precognitive dreams are the product of a faculty superior to that employed in retrospective dreams—we might hope to discover, some day, effects so abnormal in wealth of clear-cut detail that a single dream would have very high evidential value. But, if the theory of normality is right—if the faculty which dreams of the future is the *same* as the faculty which dreams of the past—we cannot expect the resemblances to the future to be any more striking than the resemblances to the past. And the latter are much less detailed than the majority of people imagine.

On the other hand, the normalist view would *allow* of the effects being far commoner than the supernormalist could permit. And it would lay down that such effects as may be observed should exhibit all grades of evidential value—from the best possible in the circumstances to the worst.

In brief, the normalist would prefer that a given value of evidence should be compounded of moderate quality and moderate frequency: the supernormalist would wish the quality to be higher and the frequency less.

What meaning, the reader may ask, do I attach to 'moderate frequency'? The answer is that it depends upon the individual. People differ enormously in the *clarity* of their dreams. A man who, in the records of fourteen nights' dreams, cannot trace more than three moderate resemblances to chronologically definite incidents of the past can hardly be expected to discover more similarly definite resemblances to the future. A man whose dreams are clearer would, presumably, discover more resemblances each way.

Judged by these normalist standards, the evidence produced in the series of experiments next to be described appears to me to vary from very good to moderate.

CHAPTER XI (*b*)

[*Note to Third Edition.*

The experiments described in this chapter were directed to ascertaining the following point:

Would the results of individual experiments, properly conducted, be likely to favour or disfavour the popular view that the faculty for precognition, assuming this to exist, is possessed by only a few abnormal individuals?

The experiment upon myself was a preliminary investigation to ascertain whether the frequency of the effect suggested normality and was high enough to render experiments on others worth making.]

The account of the following experiments, once again, is not scientific evidence, nor is it intended to be regarded as such. It is evidence for me, and part of my excuse for publication; but it is not, of course, evidence for the reader. Conviction, for him, must depend either on the convincingness of the arguments advanced in the concluding chapters, or else on the results which, according to the theory, he is likely to obtain if he makes the experiment himself—or upon both.

Personally, I found this image-hunting a fascinating and even exciting business. But it was a new kind of sport, and I made every possible blunder open to a raw beginner. Not only did I delay the attempt to recall the dream until I had been awake for half a minute or more; but I also failed to appreciate sufficiently the importance of detail in the written accounts. Incidents which should have been described in fifty words were recorded in three. The result was that, although the dreams yielded much that was suggestive of future experience, I could find, at first, little that was *identifiable* as belonging to either half of Time. There was the shower-of-sparks dream recounted in the last chapter, and five slightly more doubtful results. There was one fully described image, the original of which was seen four years later; but that was outside the prescribed limits of the test. It was not, in fact, until the eleventh day that I got the clear, conclusive[1] result I had expected.

On the afternoon of that day I was out shooting over some rough country. I was a little uncertain regarding the boundaries covered by the permission which I had obtained, and presently found myself on land where, I realized, I might have no right to be. As I crossed this, I heard two men shouting at me from different directions. They seemed, moreover, to be urging on a furiously barking dog. I made tracks for the nearest gate in the boundary wall, trying to look as if unaware of anything unusual. The shouting and barking came nearer and nearer. I walked a trifle faster, and managed to slip through the gate before the pursuers came into view. Altogether a most unpleasant episode for a sensitive individual, and one quite likely to make him dream thereof.

On reading over my records that evening, I, at first, noticed nothing; and was just going to close the book, when my eye caught, written rather more faintly, right at the end:

'*Hunted by two men and a dog.*'

And the amazing thing about it was that I had completely forgotten having had any such dream. I could not even recall having written it down.

There was nothing identifiable on the twelfth day; but the thirteenth gave another excellent result.

[1] Conclusive that my experience, in these special conditions, was opposed to the popular view referred to above.

During the day I read a novel in which one of the characters hid in a large secret loft in the roof of an old house. Later on in the story he had to fly from the house and escaped from the loft by way of a chimney.

The previous night's dream was about a large, mysterious, secret loft, which I discovered, and explored with great interest. A little later in the dream it became advisable for me to escape from the house, and I decided to do this by way of the loft.

On the fourteenth night I had four 'hot-cross-bun' results.

The net result of the experiment was that in the course of a fortnight I had been able to identify two conclusive instances of the 'effect', and six which, though not conclusive when regarded singly, could scarcely be attributed to coincidence when their number was taken into account. But the most important point was this: Not one of those instances would ever have been observed at all, had not the dreams been memorized and written down, and the records reinspected after the waking events.[1]

* * *

So far, then, the theory that the effect was merely a normal characteristic of man's general relationship to Time—but one so constituted as to elude casual observation—had been partly borne out by experiment. But, on that theory, the effect in question should be just as experimentally observable to everyone else as it was to myself. This meant that I must persuade another person to make a similar trial.

A young woman, whom I will call Miss B., good-naturedly agreed to undertake the task. I selected her mainly because she was an extremely normal individual,

[1] The number of dreams evidential of precognition was approximately equal to that of those similarly evidential of retrospection.

who had never had any sort of 'psychical' experience, and who (this was the great thing) believed that she practically never dreamed at all. Indeed, she assured me that it would be useless for her to experiment, as she had only had some six or seven dreams in the whole course of her life.

The morning after the first night she came to me and told me that it was quite hopeless. She had tried to remember her dreams the very instant she woke; but there had been nothing to remember. So I told her not to bother about looking for memories of dreams, but to endeavour instead to recollect what she had been *thinking* at the moment of waking, and, after she had got that, to try to recall *why* she had been thinking it. That worked, as I had known it would; and on each of the next six mornings she was able to remember that she had had one short dream.

Counting the experiment as starting from the first dream, she obtained, on the sixth day, the following result.

Waiting at Plymouth Station for a train, she walked up to one end of a platform and came upon a five- or six-barred gate leading on to a road. As she reached the gate a man passed on the other side, driving three brown cows. He was holding the stick out over the cows in a peculiar fashion—as if it were a fishing-rod.

In the dream, she walked up a path, she knew, and found, to her great surprise, that it ended in a five- or six-barred gate which had no business to be there. The gate was just like the one at the station, and, as she reached it, the man and the three brown cows passed on the other side, exactly as in the waking experience, the man holding out the stick fishing-rod fashion over the cows, and the whole group being arranged just like the group she saw.

The dream occurred the morning before the waking experience.

The blending of the 'past' image of the path with the 'future' image of the gate provided an excellent specimen of integration.[1]

* * *

I then asked my cousin, Miss C., to try. She was positive that she had never had any experience of this kind, and was sure that, as a general rule, she dreamed very little. She proved excellent at recovering the lost dreams, and good at noting detail. But at first she was very weak at perceiving connections, even with past events. She could not, for example, understand how a dream of walking on roofs could be connected with the experience of climbing about the roof of a bungalow with me on the previous day, though she had not been on a roof of any sort for years. She obtained, however, on the eighth day, the following first-class result:

Immediately upon her arrival at a certain country hotel she was told of a curious person staying there whom all the guests suspected, having made up their minds that she was a German. (This was during the last stages of the war.) Shortly afterwards she met this person—for the first time—in the hotel grounds. These are rather uncommon. They extend a long way, contain numbers of large, rare trees, and would certainly be taken for public gardens by anyone who did not know that they belonged to the hotel. The supposed German was dressed in a black skirt with a black-and-white striped blouse, and had her hair scraped back in a 'bun' on the top of her head.

My cousin's dream was that a German woman, dressed in a black skirt, with a black-and-white striped blouse,

[1] Miss B. had only one dream resembling a distinctive waking incident of the past within the preceding fortnight, and this dream she failed to spot until I pointed it out to her.

and having her hair scraped back in a 'bun' on the top of her head, met her in a public garden. My cousin suspected her of being a spy.

The dream occurred about two days before the event. (The record is undated, but was in my hands when the confirmatory event took place.)

She had already had one almost, but not quite, conclusive result earlier in the experiment—a dream connected with some news in a letter she subsequently received from a friend.

* * *

Mrs L., the next person to try, got an excellent result on the very first night. It related, however, to two separate experiences which occurred during the following week. The two-day limit was here exceeded; but the correspondence was so clear that the result came under the rule permitting an extension of the limit in exceptional cases.

The waking experiences concerned two public meetings at Corwen. Mrs L. went to one of these, and, in describing it to me afterwards, told me she was surprised at the large number of clergymen who seemed to have arrived out of the void to fill the building; for it did not seem to her that there was anything in the business before the meeting which could be of special interest to the Church.

She was not present at the other meeting. But my sister was there, and she *told Mrs L. of her experiences*. On putting her head in at the door she found a regular pandemonium in progress. She was about to withdraw discreetly, when the chairman, catching sight of her, called out: '*Come in, Miss Dunne, and see how we Welsh fight!*'

In Mrs L.'s dream she was at a public meeting, and was greatly annoyed by the interruptions of a clergyman in the audience, who, instead of allowing the business to proceed, insisted on preaching a sort of sermon ending in a prayer.

She expostulated. The clergyman leaned so far back that he touched her. Another man in the audience pushed against her arm. She rose, and, thumping a table, cried: 'Who is responsible for the behaviour of the audience? I know *the Welsh are notorious for bad behaviour in public*, but I will not have it here.'

Mrs L. forgot all about this dream after writing it down. Its record was not re-read by her after the second day, and so she missed it when the two meetings occurred later in the week. It was only by chance that I happened to look back through the notes and discover it.[1]

* * *

Major F., the next person approached, entered upon the experiment with considerable interest. He pointed out that, if there were anything in this business, it might mean the spotting of a Derby winner. He finished satisfied that I was perfectly right, but also satisfied, I am afraid, that the dreaming mind did not properly understand its business.

He happens to be a marine artist of considerable reputation; and on the second day of the test he set forth to paint a couple of boats which he had previously seen lying on the beach. But he found that one boat, *which was pointed at both bow and stern*, had been painted, since his last visit, in staring lifeboat (red and blue) colours. However, he made his sketch—a process necessitating, of course, long and close attention to the boat and its colours. The vessel stood on *short, green turf*. Some distance away, on a pier which came into the picture, was another long, red, somewhat boat-like object with *something draped across its middle*. Major F. took field-glasses to ascertain what this stuff was, and discovered it to be a *net*.

[1] As I have said earlier, Mrs L. had only one dream resembling any chronologically distinctive incident in the past. I questioned her on this point repeatedly during the test, as her records were voluminous, and I was puzzled by her apparent failure to get results.

The associated dream-image was that of a *red-and-blue lifeboat standing on green turf with a net draped over its middle.* This dream had occurred during the previous night.

Major F., at first, could not see the connection. He thought that the similarity ought to have extended to everything else in the dream scene, and was disappointed that this had not been the case. However, he continued the trial.

On the next day it rained heavily, and we both set out to look for a sheltered place from which to paint pictures. We entered a small house which was in course of construction, and, finding the view from the lower windows too restricted, erected a ladder against the cross-beams of the unfinished upper storey, and climbed up on to these. The ladder was a rather unusual one, in that it had square rungs.

One of Major F.'s dreams on the preceding night had been that he was climbing a ladder which did not appear to be set against any wall. It went up, so to say, into space. And it had square rungs.

He had not been up a ladder for six years.

What finally convinced him, however, was this: He dreamed that he was sailing a toy boat with a small boy protégé of his to whom he had (actually) presented this vessel. A little later on he dreamed he saw a similar boat, but full size, dismasted, and with its sails lying flat on the water. The crew were washing them. A few days after this he heard that his boy friend had been taken to a pond to sail his new boat, but instead of doing so had insisted on removing the sails, laying them flat in the water of the pond and scrubbing them.

He agreed that these three results, taken together, were conclusive.[1]

[1] I omitted to record how many of Major F.'s dream incidents appeared to relate to waking events distinctively past, but to the best of my recollection there was only one of these.

* * *

A little while before this my brother had written to me to say that he had 'got' the post-war death of General Leman, the Belgian hero, and, on opening his newspaper at breakfast, had found the announcement confronting him.

My sister, like my brother, obtained her result without the necessity of experimenting. (Both, of course, were now on the look-out for the effect.) Her evidence, however, extended into a department of science where 'Beeton' is a greater name than 'Newton'. Here, although an ignoramus, I am humble, and so I am prepared to take her word for it that the correspondence of events in this case was sufficiently detailed to put coincidence entirely out of the question.

CHAPTER XII

The situation was now a little clearer. It had been discovered that the effect was one which was apparent only to definitely directed observation, and its failure to attract general attention was, thus, sufficiently explained. But the rough-and-ready method which had been devised for the purpose of rendering it perceptible seemed to work quite well. The original hypothesis of solitary abnormality had been completely killed, and, moreover, in the light of the experiment, I did not appear to possess even a specially well-developed faculty for observing the effect. Those other people had got their decisive results more quickly than I, and, in most cases, those results had been clearer.[1]

The outcome of the experiments suggested that the number of persons who would be able to perceive the effect for themselves would be, at least, so large as to render any idea of abnormality absurd. Indeed, when one came to consider, in addition, that practically everyone has occasionally experienced that queer sense of events having 'happened before', and that most people are apt to recall suddenly an apparently forgotten dream because (there can be no other reason) something occurs which reminds them of (*i.e.*, is associated with) that dream, it became fairly clear that, if there were abnormality anywhere, it would probably pertain to those, if such there should prove to be, who were mentally debarred from observing the effect. Statistics in that respect, however, could be collected only from experiments conducted on a widespread scale consequent upon the publication of a book.

[1] My less striking results had been more numerous than those of my assistants; but, then, so had been my results similarly evidential of retrospection. There was nothing to show that I differed from the other experimenters except in a superior aptitude for 'spotting' results—*both ways.*

Meanwhile, the explanation seemed as far away as ever.

The trouble was that the effect was so extremely definite in its aspects. It was no broad, vague affair such as might be covered by some sweeping generalization (Relativity, for example, or a two-dimensional theory of Time); it bristled with peculiarities; it presented clues which pointed like signposts to half a dozen solutions—mostly contradictory. And, though it was easy to devise explanations which should cover some of the facts, it was difficult to find anything which could fit them all.

In the hope of obtaining additional data, I recommenced experimenting upon myself, the immediate object being to ascertain whether there were any observable differences between the images which related to the future and those which related to the past. As it turned out, the most careful observation failed to bring to view any such distinguishing features.

In the course of these further experiments, however, I came upon three dreams of a specially illuminative kind, and these, perhaps, had best briefly be described.

The first afforded a fairly clear example of an associational chain running from 'past' to 'future'. The connecting link was the idea of *spilled ink*, which idea entered into both the related waking experiences.

Waking experience (1): *before the dream.*—Watched a friend seated at a table filling a fountain-pen, and thought he was going to spill the ink.

Waking experience (2): *after the dream.*—Read a French detective story. The detective seemed to be unusually incompetent, and, towards the end of the book, I began to wonder when he was going to exhibit some sign of the skill with which the reader had been asked to credit him. In the *dénouement* he pretended to stumble, and, in so doing, upset some ink over a table at which the villain was seated. The latter, to save his clothes, threw himself back in his chair, raising his hands above the flood. Whereupon, the detective seized one hand and slapped it down first into the ink and then on to a piece of blotting-paper, thus obtaining a set of finger-prints. He then triumphantly denounced the criminal.

Dream: between the two waking experiences.—A famous detective was going to give us an exhibition of his skill. We waited a long time, but he seemed quite incompetent. Finally, he pretended to stumble, and, in so doing, spilled ink *from a fountain-pen* over the criminal, whom he then triumphantly denounced.

The second dream exhibited a similar associational chain, but in this case the link—*shooting dangerous game with a revolver*—was much clearer.

Waking experience (1): *before the dream.*—Saw pictures of a lion-shooting expedition. My brother was thinking, at the time, of joining such an expedition, and I began to wonder what guns he ought to take. While considering the merits and demerits of various weapons, I was reminded of an enormous seven-chambered revolver I had seen in a Paris gun shop, which apparatus was supposed to be part of the equipment of any up-to-date hunter of lions. I wondered, with some amusement, what lion-shooting with a revolver would be like.

Waking experience (2): *after the dream.*—Read Ethel Sidgwick's *Hatchways*. Two chapters are devoted to the episode of a leopard, which has *escaped from a menagerie*. It has appeared near a country house where a sort of children's school treat is in progress, and *has killed a goat*. Later on, the hero is saved from the animal by a retired explorer, who arrives in the nick of time and kills the beast with two shots from a borrowed revolver.

Dream: between the two waking experiences.—Looking from the windows of a country house, saw the head and shoulders of a *lion* moving through a cornfield. It was known in the neighbourhood that this lion had *escaped from a menagerie*, and that it *had killed a goat*. Wondered if I could hit it from the window with my revolver, but decided that the range was too great. Decided to lie up alongside the track in the cornfield, and to wait till the beast repassed. Felt, however, that I should prefer to be armed with something better than a revolver. Went out to try to get a rifle.

The third dream provided an example of a perfect integration, the component parts of which were related to impressions received before and after the dream.

Waking experience (1): *before the dream.*—Saw in the garden of an hotel where I was staying the bottom, *minus* the sides, of an old, small, flat-bottomed boat.

Waking experience (2): *after the dream.*—My sister persuaded me to go with her to one of the Olympia motorcycle shows, as she wanted my opinion on a small 'scooter' which had caught her fancy. It was a neat-looking little thing called the 'Unibus', and it was entirely different from the other scooters in the show, inasmuch as it was built on motor-car principles, with shaft, gear-box, etc. It was equipped with a little seat of curious shape (on all scooters that we had seen hitherto, one stood on the baseboard). Also, it was fitted with a shield for the protection of ladies' dresses. I pointed out the advantages of this last feature, and added that in ordinary scooters she would get her feet horribly wet and muddy. As I said that, there flashed through my mind the old curious conviction: This has happened before. Knowing what that meant, I set to work and presently revived the lost memory. It belonged to a dream, and what was more, a dream which I had recorded. On my return home I looked up the notes, and found that they had been made *two years before*.

Dream: between the two waking experiences.—Saw my sister coming down a street, sitting in an extremely curious little motor-car. (I had made a sketch of this machine, which was simply the 'Unibus' without its shield.) Called out to her something about getting her feet wet. Saw water in the roadway up to the level of the low, oval platform.

The notes stated that the platform of this tiny car was *the piece of a flat-bottomed boat* I had seen nine or ten days before.

Since we have got on to the subject of long-range association with a dream in the middle, I may as well describe the most perfect example of the kind I have ever experienced. The gap between dream and future event was about twenty years.

Waking experience (1): *before the dream.*—When a small boy, between twelve and fourteen, I read with enormous interest Jules Verne's *Clipper of the Clouds*. Readers of that book will probably remember the illustrations of the author's idea of a flying machine. These showed a long, dark hull of about the size and shape of a modern 'Destroyer', except that it had a ram bow. This thing, which looked as if it had got off the sea and into the air by mistake, was supported solely by a cloud of tiny screw propellers mounted on a forest of thin metal masts. There were no wings, or anything of that sort.

Waking experience (2): *after the dream.*—Some twenty years later, in 1910, I made the first decisive flight in the first aeroplane which possessed complete inherent stability.[1] It was a rather exciting episode. The thing got off too soon, bounced—and, when I recovered my scattered wits, I found it roaring away over the aerodrome boundary, climbing evenly, and steady as a rock. So I left well alone,

[1] Mr. L. Gibbs had previously persuaded a similar machine of my design to leave the ground; but the flight on that occasion was limited to a few yards.

and allowed it to look after itself. This it did till the engine gave out (usually a matter of three minutes in those days). The sensation was most extraordinary. The machine, like all those of my design, was tailless, and shaped, as viewed from below, like a broad arrow-head *minus* the shaft. It travelled point foremost, and, at that point, there was fitted a structure like an open (undecked) canoe, made of white canvas stretched over a light wooden framework. Seated idly in this, and looking down over the sides at the cattle scampering wildly around three hundred feet below, the whole of the main structure of the aeroplane was away back behind the field of vision, and the effect produced was that one was travelling through the void in a simple open canoe.

Dream: between the two waking experiences.—A few days after I had read, as a small boy, Jules Verne's book, I dreamed that I had invented a flying machine, and was travelling through space therein. It must be borne in mind that I had never heard of, or conceived the possibility of, any flying machine different from the great metallic, screw-supported 'clipper of the clouds'. Yet in my dream I was seated in a *tiny open boat constructed of some whitish material on a wooden framework.* I was doing no steering. And there was no sign of anything supporting the boat.

I may add here that the boat-like *nacelle* of the 'Dunne' biplane had not been added on account of any lingering, unrecognized memory of the dream. The earlier machines had no such feature. This had been attached as an afterthought, simply in order to reduce the 'head-resistance' of the pilot, which resistance, at that particular place, was believed to exercise a detrimental effect upon the stability of the apparatus.

I never forgot that dream, and recalled it with amusement when, in 1901, being on sick-leave from the Boer War, I set to work in earnest to devise some 'heavier-than-air' contrivance, which should solve the great military problem of reconnaissance. But it seemed to me a dream natural enough for a boy, and I did not then perceive the significance of the *appearance* of the dream-machine—indeed, I could not do so, for the related constructional development did not come till ten years later. By then I had dismissed the dream as of no importance, and it was only recently that I realized that the corroborative detail of the little, white, open boat classified the whole as an anticipation of future experience.

* * *

Granted that the dreaming attention ranges about the associational network without paying heed to any particular 'present', there is nothing astonishing in its lighting on an image many years 'ahead'. This, in fact, is exactly what we should expect, for in its 'backward' travel it often lights on images many years 'behind'.

But, when it comes to computing the proportion which the images of the past bear to the images of the future, in a given series of dreams, one is apt to be misled. For the images which relate to events a long way behind can be recognized and counted; but those which relate to events similar distances ahead cannot be identified. Hence, the only way to strike a balance is to confine the statistics to the range of a few days either way. Images which relate equally well to either past or future—such as those of friends, and of everyday scenes—should not be counted. Images which are apparently of the past should be submitted to the same severe scrutiny as are those which are apparently of the future, for coincidence will operate just as effectively in either direction.

Computing in this fashion I have found that the images which relate indisputably to the near-by future are about equal in number to those which pertain *similarly indisputably* to the near-by past.

* * *

[*Note to Third Edition.*

The paragraph italicized above was written seven years ago, and I marvel that I did not realize then that I had sketched the outlines of a statistical experiment far more convincing and immensely simpler than the one previously described. It would require to be conducted on a far larger scale; but, that done, it would provide much better evidence of the probable distribution among individuals of any precognitive faculty that might be presumed to exist. Moreover, if the scale were sufficiently large, it might even produce irrefutable scientific proof of the *fact* of precognition. It was not, however, until 1932 that all this dawned upon me. Then I conducted promptly a small-scale experiment upon these lines. The results were overwhelmingly in favour of the new dream theory. They are given in the Appendix to this book.]

CHAPTER XIII

Why only in dreams? That was the question which blocked all progress. Every solution which could reduce Time to something wholly present ruled that the pre-images should be just as observable when one was awake as they were when one slept. So, why only in dreams?

I should be ashamed to confess how long a period elapsed before I saw that, in framing that question, I was *begging the question.* The moment, however, that I did realize this, I proceeded to put the matter to the test.

A little consideration suggested that the simplest way to set about a waking experiment would be to take a book which one intended to read within the next few minutes, think determinedly of the title—so as to begin with an idea which should have associational links with whatever one might come upon in that future reading—and then wait for odds and ends of images to come into the mind by simple association.

Obviously, one could save a lot of time by rejecting at once all images which one recognized as pertaining to the past. Also, since the images would be perceived while awake and with one's wits about one, one might rely more upon one's memories of them than one could when the memories were formed sleeping, and thus save a vast amount of writing. A brief note of each image should suffice.

The first experiment was a gorgeous success—until I discovered that I had read the book before.

It was interesting, however, as showing the tremendous difficulty the waking mind experiences in freeing itself from its memories. I spent by far the greater part of the time in rejecting images of the past and starting afresh with a mind comparatively blank.

Apart from the items which related to the book (already read), I got only a few ideas, mostly concerning London and the exterior and interior of clubs. The only exception was the single word '*woodknife*', which drifted into my mind, seemingly, from nowhere. A little reflection satisfied me that I had never in my life come upon such a word, so I jotted it down.

Two or three days after this I moved, quite unexpectedly, to London. On my arrival, I went to my club, and having for the moment nothing better to do, proceeded to the library, picked out a newly published novel, and tried a second experiment. Result—nil. In fifteen minutes I got only eight images, which did not clearly belong to the 'past' half of the associational network. One of these eight related to a *kangaroo hunt* in Australia—riders and hounds chasing pell-mell after the leaping animal. Another comprised the single word '*narwhal*'. There was nothing in the book that fitted, and presently I threw it aside.

I then drifted into a little inner library, which is an excellent place for a nap. I chose a comfortable armchair, and, for appearances' sake, equipped myself with another volume—R. F. Burton's *Book of the Sword*, opening this in the middle.

Immediately my eyes fell upon a little picture of an ancient dagger, underneath which was inscribed '*Knife (wood)*'. I sat up at that, and began to dip into the book, turning back after a moment to page 11. There I came upon a reference to the horn of the narwhal. Reading on, I found on the succeeding page the words, '*The "old man" kangaroo, with the long nail of the powerful hind leg, has opened the stomach of many a staunch hound.*'

Now, there was nothing conclusive here, but it was just the sort of suggestive but uncertain thing one keeps on getting throughout the dream experiment, while one is waiting for one's decisive result. I was, therefore, encouraged to proceed.

I tried next with Baroness von Hutten's book, *Julia.* Result—a quarter of a sheet of notepaper of material, the only thing that fitted being '*pink house*', there being a reference in the book to '*pink houses*'. (Not good enough.)

Arnold Bennett's *Riceyman Steps* served for the next experiment. I got only three lines of material, but these contained the words, '*I am entitled to say*'. On opening the book I found in the first paragraph the words, '*The man himself was clearly entitled to say*'.

Then I tried with Mason's *House of the Arrow.* Here I altered the procedure. I opened the book at the beginning, and found the name of one of the characters, being careful not to glance at any other page. It seemed to me that a name which would be likely to occur in close connection with many of the incidents of the story would provide a better associational link than does the mere idea of the book's title.

I do not know if the present reader is acquainted with the *House of the Arrow*, and, if he is not, I am most unwilling to spoil for him, even in the interests of science, the enjoyment of a first-class detective story. So I will merely say that the centre knot of the whole tangle—the thing upon which everything in the plot hangs—*is a clock pointing to half-past ten.* This feature, however, does not come into the story till halfway through the book.

The character I had chosen from the opening pages as an associational link accompanied the detective throughout the latter's investigation. Concentrating attention on that character, the first image I saw and noted was that of *a clock pointing to half-past ten.*

With Lord Dunsany's book, *The King of Elfland's Daughter*, I got '*Long cliffs of crystal looking over dark sea.*

Fireflies dancing over this sea'. Not a bad description of the night scene pictured in the book, where the long crystal cliffs look down upon a mist-covered plain over which the lights of Elfland are dancing, advancing, and receding.

I then tried a book of Snaith's, taking the heroine's name as an associational link. Here I failed completely. But, in the middle of this experiment, I got one very curious image.

It was that of an umbrella with a perfectly plain, straight handle, a mere thin extension of the main stick, and of much the same appearance and dimensions as the portion which projected at the ferrule end. This umbrella, folded, was standing unsupported, *upside down*, *handle on the pavement*, just outside the Piccadilly Hotel.

I happened to pass that way in a bus next day. Shortly before we got to the hotel I caught sight of a most eccentric-looking figure walking along the pavement in the same direction, and on the hotel side of the street. It was an old lady, dressed in a freakish, very early-Victorian, black costume, poke bonnet and all. She carried an umbrella in which the handle was merely a plain, thin, unpolished extension of the main stick, of much the same appearance and dimensions as the portion which projected at the ferrule end. She was using this umbrella—closed, of course—as a walking-stick, grasping it pilgrim's-staff fashion. But she had it *upside down*. She was holding it by the ferrule end, and was pounding along towards the hotel with *the handle on the pavement*.

I need hardly say that I had never before in all my life seen anyone use an umbrella that way.

* * *

These experiments showed me that, provided one were able to steady one's attention to the task, one could observe the 'effect' just as readily when awake as when sleeping.

eadying of attention is no easy matter. It is : makes no call upon any special faculty, but ıand a great deal of practice in controlling the n. Hence, to anyone who is desirous merely of satisfying himself as to the existence of the 'effect', I should recommend the dream-recording experiment in preference to the waking attempt.

But, for *studying* the problem, the waking experiment is of distinct value, because one can follow a great deal of what one's mind is doing. Also, there is no dream-story to complicate matters.

In my own case, I employed this experiment mainly in order to seek for the *barrier*, if any, which divides our knowledge of the past from our knowledge of the future. And the odd thing was that there did not seem to be any such barrier at all. One had merely to arrest all obvious thinking of the past, and the future would become apparent in disconnected flashes. (For, however difficult and troublesome the process, that was what, ultimately, it resolved itself into.) Yet, if one tried to follow up the 'memory train' from past to future, one came, not so much to a resisting barrier, but to an absolute blank. Moreover (and this I discovered by separate experiment), if one allowed the attention to pass from the image under consideration to another which was *manifestly* associated therewith, one remained, so to say, in the 'past' part of the network. There, attention was completely at home. The associated images followed one another in swift, easy succession; attention ran on and on without noticeable effort or fatigue.

It was only by rejecting manifest associations with the last image, and waiting till something apparently *disconnected* took its place, that attention was enabled to slip over the dividing line.

CHAPTER XIV

There remains one more dream to be described. While not, perhaps, completely conclusive, it was so nearly so that it had to be taken into serious account. And since, if it really did relate to the future, it could not possibly fit in with the solution I happened to be favouring at the time, it caused me to abandon work on those particular lines, and to hark back to an earlier theory. And this, as it turned out, was wholly fortunate.

On the morning after the dream I was, while dressing, engaged in following up a long train of reminiscences of my school days—a train which led, in perfectly logical sequence, to the memory of an adventure with a wasp. As a boy I was terrified of these insects, and could hardly bring myself to remain in the same room with one. Imagine my horror, then, when, during a meal in a room with an open window, a large wasp entered, flew to me, settled on my neck and proceeded to crawl round deep down inside my Eton collar. I sat there, white as the table-cloth, while a master adjured me, quite unnecessarily, not to move. To this day I can remember the horrid sensation of the insect's soft, faintly felt perambulations. And so, forty-four years later, on this particular morning, when my train of thought had brought me to that early memory, I tried to recall the feel of those crawling feet. As I did so I happened to be combing my hair; the comb caught at a particular place on the crown of my head, and instantly there came back to my mind a dream of the previous night. I had dreamed of that feeling of something catching in my hair at that precise point of my scalp, had been convinced that *a wasp was crawling there*, and had called to a companion to take it off.

Now, assuming that this was an anticipatory dream—an instance of the 'effect'—we have the following facts to consider.

The simultaneous presentation to consciousness of the sensory impression of the comb in the scalp and the memory image of the feel of the wasp's feet, was a straightforward enough example of the process of forming an association by 'contiguity'. And, *before that association had been formed*, it was presented in the dream in the shape of an integration.

A very pretty mixture of experience.

PART IV

TEMPORAL ENDURANCE AND TEMPORAL FLOW

CHAPTER XV

Before we begin to look for an explanation, it might be as well for us to glance, briefly, at what precisely it is that we have to explain.

First, of course, there is the 'effect' itself—the apparent temporal disorder of the presentations. The actual order of experience, such as might be recorded in a diary, runs thus:

a' a pre-presentation of A.
a'' a re-presentation (memory) of a'.
A a presentation.
a a re-presentation of A.

If we accept the evidence afforded by the dream described in the last chapter, the matter becomes more complicated in this respect: It looks as if A, in the above list, might be *any* sort of compound of presentations.

Next, we have the following to consider:

As the result of observing an image of future experience, the experimenter takes pencil and paper, and notes down, or even makes a sketch of, the details of the pre-image observed. In so doing, he is performing a definite physical act. *But it is an act which would never have been performed had he not observed that pre-image.* In other words, he interferes with that particular sequence of mechanical events which we postulate as the backbone of our 'conscious automaton' or materialistic theories.

This is barefaced 'intervention'. But it implies something more. *These future events are, at any rate, real enough to be experienced as pre-presentation; yet—since, as we have just seen, the observer can alter his course of action as the result of his pre-observation—they are events which, theoretically, may be prevented from happening. Are we, then, to say that they are only partly real—less real, for instance, than are past events? That is another question our explanation has to answer.*[1]

Furthermore, this ability of the observer to interfere with the course of brain events introduces the question of 'free-will'. Our solution will have to make a satisfactory statement in that connection.

Finally, it is essential that the explanation does not contradict the already known facts of psychology and psychophysics. And of those facts there are some which greatly limit our range of permissible speculation. On the psychical side we have the fact—dwelt upon in Chapter XIII—that the memory '*train*' does not run through into the 'future'. It ends in the 'present'. On the psychophysical side we have all that is included in the usual evidence for parallelism, and, in particular, the known fact that concussion of the brain apparently destroys or paralyses recently formed memories. There can be no question but that here something more than a mere 'motor habit' is affected: the patient's mind with regard to such immediately previous events seems to be a complete psychological blank.

[1] This, of course, is the classical objection to the notion of prophecy.

CHAPTER XVI

It is worth noting that Relativity admits of 'seeing ahead' in Time, in the sense that what is future to Jones may be present to Brown. But it does not admit of an event in the remoter future of Jones appearing to Jones a day or two before an event in his nearer future. And that is our problem.

* * *

It must be borne in mind that material records are indications of the *past* only, so far as the thing on which those records are imprinted is concerned. If, on inspecting a target at a given instant, you perceive a round, punctured hole in the corner, you may infer that a bullet has passed through at that point. But nowhere does that target offer you any indication that another puncture is presently going to appear therein—at, say, half an inch from the centre of the bull's-eye. It is true that, from a complete knowledge of all the mechanical movements which were going on in that quarter of the universe at the moment of your inspection, you might, if you were possessed of some sort of superlative intelligence, be able to deduce that a bullet would shortly strike the bull's-eye at the point in question. But that is to confuse the issue. It is to introduce a host of indications *external to* the one we are considering—which one is the state of the target. That state offers no indication concerning the coming puncture. So uncommunicative is it that, in working out your prophecy, you would leave the question of present damage or lack of damage entirely out of consideration: it could not affect your decision. The *target* contains no 'record' of its own future—the indications you use are, in fact, everywhere *except* on that surface. But the punctured corner of the target is a record of the past history of the target; and it is from that record, and not from a knowledge of what exactly was going on throughout the whole of that quarter of the universe at some earlier moment of Time, that you deduce the past impact of the bullet.

Punctures in the target are indications of the future, in the sense that they are evidence of the directions which the bullets may be taking, and so indications of what may be going to happen to the stop-butt at the back of the target; but they are not indications of future punctures in the target itself.

Now, the brain is a material organ, and the state of the brain at any given instant is no more an indication of what the world outside the brain is going to present to that brain in the future than is the state of our target an indication of where the next bullet is going to strike, or whether a new one is going to strike it at all.

CHAPTER XVII

It is never entirely safe to laugh at the metaphysics of the 'man-in-the-street'. Basic ideas which have become enshrined in popular language cannot be wholly foolish or unwarranted. For that sort of canonization must mean, at least, that the notions in question have stood the test of numerous centuries and have been accorded unhesitating acceptance wherever speech has made its way.

Moreover, the man-in-the-street is, all said and done, *Homo sapiens*—and the original discoverer of Time. It was from him, and from him alone, that science obtained that view of existence.

His conclusions regarding the character of his discovery seem to have been very emphatic in detail, if slightly uncertain in synthesis. His idea was that temporal happenings involved *motion in a fourth dimension.*

Of course he did not *call* it a fourth dimension—his vocabulary hardly admitted of that—but he was entirely convinced:

1. That Time had length, divisible into 'past' and 'future'.

2. That this length was not extended in any Space that he knew of. It stretched neither north-and-south, nor east-and-west, nor up-and-down, but in a direction different from any of those three—that is to say, in a fourth direction.

3. That neither the past nor the future was observable. All observable phenomena lay in a field situated at a unique 'instant' in the Time length—an instant dividing the past from the future—which instant he called 'the present'.

4. That this 'present' field of observation *moved* in some queer fashion along the Time length; so that events which were at first in the future became present and then past. The past was thus constantly growing. This motion he called the 'passage' of Time.

There is a point here worth noting—a point which we shall have to discuss more fully later on. An examination of the last paragraph will show that many of the words therein refer to *another* Time, and not to the Time stretch over which the passage of the 'present' field of observation was supposed to take place. This, perhaps, will be more readily seen if the paragraph be repeated with the words in question italicized.

4. That this 'present' field of observation *moved* in some queer fashion along the Time length; so that events which *were at first* in the future *became* present and *then* past. The past was thus *constantly growing*.

The employment of these references to a sort of Time behind Time is the legitimate consequence of having started with the hypothesis of a *movement* through Time's length. For motion in Time must be timeable. If the moving element is everywhere along the Time length at once, it is not moving. But the Time which times that movement is another Time. And the 'passage' of that Time must be timeable by a third Time. And so on *ad infinitum.*[1] It is pretty certain that it was because he had a vague glimpse of this endless array of Times, one, so to say, embracing the other, that our discoverer abandoned further analysis.

But he adhered to his two main conceptions—the Time length and the Time motion. And he coined special phrases with which to convey to his entirely comprehending companions those two very practical and useful

[1] This, of course, has been pointed out before now—as an objection to the Newtonian idea of a Time which flows.

ideas. He spoke of a 'long' Time and a 'short' Time (never of a broad or narrow Time). He referred to the 'remote' past and the 'near' future. He said, 'when to-morrow comes', and, 'when I get to such and such an age'. In his more poetical moods, he declared that Time 'flew', and that the years 'rolled by': he wrote of 'life's journey', and of living 'from day to day'.

He symbolized this general conception of Time in several ways; most exhaustively, perhaps, in his sheets of piano music. In these, the dimension running up-and-down the page represented Space, and intervals measured that way represented distances along the instrument's keyboard; while the dimension running across the page from side to side represented the Time length, and intervals measured that way indicated the durations of the notes and of the pauses between them. But that did not complete the symbol. So far, the page represented merely what we should, to-day, call a 'Space-time continuum'. In order to complete the symbol, it was intended that the player's point of vision should *travel* from left to right along the model Time dimension, and that the written chords should be played as this moving point, representing the moving 'present', reached them.

In another case the Time dimension was represented by the circumference of a circle, this length being marked off into portions representing Time distances. But that alone did not suffice to convey his conception of Time. There was no moving 'present'. So he added a pointer to represent this 'present', and set it moving over the symbolical Time dimension by means of machinery. The entire contraption was then not only a symbol, but an actual working model of Time as he conceived it. It was an extremely useful device; and he called it a 'clock'.

Now, a clock-face without hands; a sheet of music which directs that all the chords are to be played with one resounding crash; and the concept of a Time length in which every part is equally present to a seventy-year-long observer: these three things are, to the man-in-the-street, exactly equivalent in value.

For he did not conceive Time as having length (or infer that Time had length) save for some very good and quite imperative reason. Nor is that reason in any way hidden or obscure. We all perceive phenomena as being arranged in two sorts of order. There are those which appear to be merely separated in Space, and those which appear to be '*successive*'. That difference is 'given'; it is there; it confronts us, do what we will, or think how we may. We must have conceived or perceived that Time had length merely as part and parcel of an attempt to account for this apparent *succession* of phenomena. So it would have been equally part and parcel of that attempt that the Time length should be regarded as a length-moved-over, a dimension in which we travelled from second to second, from hour to hour, from year to year, thus coming upon the Time-separated events one after the other, just as we come upon objects in our mundane journeys. The original concept must have appeared as a single one—that of length-moved-over. That the two component ideas in this complex—Time length and Time movement—may possess any analytical value regarded *entirely* apart from each other demands a considerably more advanced power of reasoning.

It was not until comparatively recent years that it seems to have occurred to anyone that the man-in-the-street's imagined, but unchristened, fourth dimension might prove to be a 'real' fourth dimension, akin to any of the three dimensions of Space. D'Alembert (1754) wrote of a friend of his who had conceived this notion.[1]

[1] I am indebted to Mr Edwin Slosson for this piece of information. Professor Fritz Paneth tells me that Fechner, writing under the name of

But the earliest treatise on the subject that I have read is a monograph by C. H. Hinton entitled *What is the Fourth Dimension?* and published in 1887.

[Hinton described a little model system of lines nearly upright but sloping in different directions and supposedly all connected to a rigid framework. If this framework with its fixed, slanting lines were to be passed slowly downward through a horizontal fluid plane which stretched at right angles to the direction of the motion, 'there would be the appearance of a multitude of moving points in the plane, equal in number to the number of straight lines in the system'. If solid threads of matter were substituted for the lines, these moving points (cross-sections of the threads) would appear as moving atoms of matter to an imagined two-dimensional being inhabiting the fluid plane and regarding it as all the Space there was. Similar considerations would hold good for an arrangement of four-dimensional threads of matter passing through three-dimensional Space. 'Were such a thought adopted, we should have to imagine some stupendous whole, wherein all that *has ever come into being or will come co-exists*, which, passing slowly on, leaves in this flickering consciousness of ours, limited to a *narrow space* and a *single moment*, a tumultuous record of changes and vicissitudes that are but to us.' The italics are mine.]

Readers who are not used to visualizing geometrical figures may find Hinton's description a little difficult to follow. It might be as well, therefore, to present the idea in a rather simpler form, and to illustrate this by

Dr Mises, published in his *Vier Paradoxe* (1846) an account of Time as a fourth dimension which forestalled Hinton's (mentioned hereafter) and contained a diagram more like that on page 138.

The notion of an 'Everlasting Now' in which past, present and future co-exist is, I believe, one of the commonplaces of Oriental philosophy.

means of a diagram. But, before we do so, a word or two of explanation regarding Time diagrams in general may not come amiss.

A dimension is not a line. It is any *way* in which a thing can be measured that is *entirely different* from all other ways. In geometry we are measuring a fundamental thing called '*Extension*'—a thing which is simply the formal opposite to nothingness. We find that, if we set about measuring this in ways which appear to be each totally different from all the others, these ways must appear to be each at *right angles* to all the others. Thus, if we choose to start by regarding north-and-south as one way (one dimension), we may consider east-and-west as another way, because we can measure off distances east-and-west without ever moving northward or southward at all. A third way in which we could measure without infringing on the other two ways is up-and-down. If Time has length—which is extension—then Time provides us with a fourth way, for we could measure along Time without moving in any of the dimensions already mentioned. A fifth way . . . but we have, as yet, no *names* for any other ways. Yet, theoretically, there may be an unlimited number of such ways, each at right angles to all the others. Mathematicians think nothing of considering ten of them. But we cannot *visualize* more than three at a time, because our bodies and brains are machines which are not constructed to work in more than three dimensions.

When it comes to drawing diagrams, we find ourselves limited to the use of the two dimensions in which the paper is extended—*viz.*, up-and-down and from-side-to-side. But we may use these two dimensions to represent *any* two dimensions we please—the fourth and the fifth, for example, or the first and an imagined one-hundredth—because, whichever two dimensions we choose to represent, these must be at right angles to each other in exactly the same way as are the dimensions of the paper. Thus, we can say that one dimension of the paper represents Time, and the other a dimension of Space, and then draw diagrams exhibiting the relation of real Time to this Space dimension. For, if Time is really extended (has length), it would be possible for the diagram to be placed, in exactly that fashion, in a plane which extended one way in Time and the other way in Space.

But what about the remaining two dimensions of Space? Well, one of them may be considered as standing out at right angles to the plane of the paper and may even, if you like, be shown in a perspective view. The other cannot be shown at all, or even imagined. You merely know that it must be *considered* as extending at right angles to the other three. But the simpler kinds of Time diagram deal with problems in which the consideration of more than one or two dimensions of Space is unnecessary.

In the present diagram, we shall consider the side-to-side dimension of the paper as representing Time, and the up-and-down dimension as representing Space. In order to avoid all chance of any reader confusing a dimension with a line, I propose to place a little dimension-indicator in the corner of the picture, just as a cartographer places in the corner of his map a little diagram showing the points of the compass. Time will be indicated by T, and Space by S.

Here, then, is Hinton's idea, pictured in two dimensions, but with lines of a rather more varied character than had those which he took into consideration, and with the whole model arranged to work horizontally instead of vertically.

The full lines represent material threads extending (enduring) in Time. If you examine any one of these lines, you will notice that the points of which it is composed are placed at different positions in Space (different heights on the page) at different moments in Time (different distances from the margin). The dotted line *AB* represents a section of what Hinton called a 'fluid plane' (you may imagine the rest of it as sticking out at

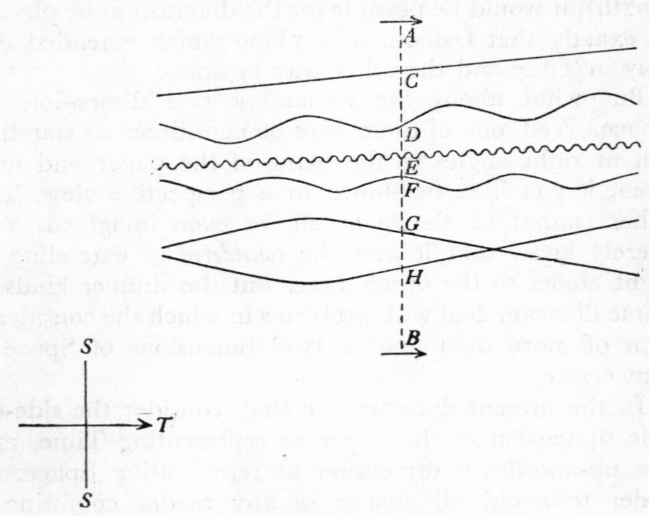

Fig. 1.

right angles to the paper, though that is quite unnecessary). The arrow-head to the *T* in the little dimension-indicator shows that *AB* is to be regarded as moving, without tilting one way or another, straight along the Time dimension. The arrows at the top and bottom of the moving line are merely there to reinforce this idea. They will be omitted, as a rule, in subsequent diagrams.

If *AB* were to travel thus, the little bits of the full lines, where these are intersected at *C*, *D*, *E*, *F*, *G*, and *H*, would appear as moving either towards *A* or towards B—as moving, that is to say, in Space. (If you will cut, in a piece of paper, a fine slit to represent *AB*, lay this on the diagram with the slit parallel to *AB*, and then slide the paper in the direction of the arrow, you will see these apparent movements with great clearness.)

A creature whose field of observation was thus limited to *AB* would be aware, therefore, of a little world of moving particles. But you and I, whose field of observation covers the whole diagram, perceive that the *actual bits* of the full lines intersected do not really move about on the page: what happens is merely that the sectional *views* of the lines move as our eyes follow the movement of *AB*. And the only thing which seems to us really to move over the page is the line *AB*.

So, according to Hinton's theory, a being who could see Time's extension as well as that of Space would regard the particles of our three-dimensional world as merely sectional views of fixed material threads extending in a fourth dimension, and would consider that the only thing in the entire Cosmos that really moved was that three-dimensional field of observation which we call the 'present moment'.

Hinton assumes thus that the past and the future 'co-exist', and that our experience of change is due to a relative motion between this Time extension and that 'narrow space and a single moment' which is the present. But he refrains from noting that such relative motion must *take Time.*

As a contribution to the subject, Hinton's exposition was remarkable, in that it clearly indicated the part that must be played by matter in any careful interpretation of the man-in-the-street's vague idea. According to Hinton, matter, as exemplified by his 'threads', *extended* in the Time dimension.

The man-in-the-street has never definitely carried his analysis thus far. To him, it seems essential that something should move in Time; but there is no evidence that he has ever realized that there would be a vast difference between (*a*) a system in which his three-dimensional field of observation moved through a stationary world of four-dimensional matter, and (*b*) a system in which he and a three-dimensional material world moved together, *en bloc*, through a blank.

The latter concept is, of course, entirely devoid of meaning. Its acceptance may, in fact, be said to constitute the great Time Fallacy, Movement of the universe *as a whole* through a thousand such featureless dimensions could not make the slightest difference to what was going on in that universe: it could not explain or account for any phenomenon whatsoever, temporal or otherwise. There would be no change, no experience of succession, which would not be equally apparent in the absence of that supposed motion. *Nor would the concept of such a motion amplify or abstract from any concepts that you can entertain without thinking of such a motion.*

The man who allows himself to drift unwittingly from his original concept of an occupied Time—a dimension in which he travels from event to event—and who begins to entertain in its place the meaningless idea of his travel through an empty and ineffectual continuum, seldom proceeds very far in his thinking before, perceiving the nonsensicalness of his new idea, he decides that 'there is no such thing as Time'.

CHAPTER XVIII

To Hinton there was no qualitative difference between the Time dimension and the dimensions of Space. He *started* with four dimensions of Extension, all fundamentally alike, and his problem was to discover why any human being should regard one of these as specially distinguished from all the others. He found his answer in the idea of a three-dimensional field of observation moving up the four-dimensional block. This, it will be noted, made the apparent Time dimension the same for all observers, no matter which way the bundles of material threads representing their bodies happened to be inclined in the whole dimensional extension. His travelling field would, thus, be a constituent of the universe which existed independently of the existence of any individual observer.

Mr H. G. Wells took a slightly different view. And, in *The Time Machine*, published seven years later, he, through the mouth of one of his fictional characters, stated his case with a clearness and conciseness which has rarely, if ever, been surpassed.

He begins by insisting on the *necessity* of regarding Time as a fourth dimension. (Hinton had not perceived this.) It is a way in which matter *must* be measured.

'There can be no such thing as an instantaneous cube . . . any real body must have Length, Breadth, Thickness, and . . . Duration.'

Matter, thus, for him, as for Hinton, extends (endures) in Time.

'For instance, here is a portrait of a man at eight years old, another at fifteen, another at seventeen, another at twenty-three, and so on. All these are evidently sections, as it were, Three-dimensional representations of his Four-dimensional being, which is a fixed and unalterable thing.'

(The portraits in question would have needed to be sculptured, three-dimensional figures. But the meaning is clear.)

He emphasized and re-emphasized the fact that there was no qualitative difference between a Time dimension and a Space dimension. There was an apparent distinction, drawn by the observer, but no such distinction if you left the observer out of it.

'There are really four dimensions, three of which we call the three planes of Space, and a fourth, Time. There is, however, a tendency to draw an unreal distinction between the former and the latter, because it happens that our consciousness moves intermittently in one direction along the latter from the beginning to the end of our lives.'

A little later on, he refers to the Time-moving elements as 'our mental existences'. Note the use of the plural. There is no all-embracing moving *stratum*, filling Space between the different observers, but a number of 'mental existences', one for each observer, and it is the motion of these which alone determines which dimension is Time.[1]

Now, that statement implies something which Wells did not specifically mention. Each of such mental existences would be centred in or about the corresponding observer's brain, and so, in its travel, would be bound to follow whatever bundle of fixed lines in the four-dimensional extension represented that brain. Hence, if it were the travel of the 'mental existence' which caused

[1] In the story which follows, the hero is granted an amount of geometrical freedom considerably greater than such a theory would allow. But that—to the reader—is a matter for rejoicing rather than complaint.

the observer to make an artificial distinction between Time and Space, each observer would regard Time as *stretching in the direction in which his body line extended.* It would follow that his body line would seem to him to be running *straight* up this Time dimension of his, and not to be bending this way and that in Space—*i.e.*, sitting in a railway train, he would seem to himself (until he began to speculate about it) to be at rest.

Moreover, the body lines of different observers are never parallel. Our bodies do not remain a constant distance apart from one another in Space. Therefore, different observers would hold slightly differing opinions as to the correct directions of the Time and Space dimensions.

For the rest, we may note that, like Hinton, Wells fails to mention that anything which moves in Time must *take Time* over its movement.

* * *

The Relativists exactly reversed the procedures of the nineteenth-century Time-dimensionalists. Certain apparent anomalies in certain optical experiments led Einstein to enunciate, for the first time in history, not merely the idea that different individuals could hold different views regarding both Time (as told by clocks) and Space (as measured by rods), but that such judgements would be equally valid. From this, Minkowski deduced the existence of a four-dimensional extension in which there was no qualitative distinction between the dimensions, but only an apparent distinction, each observer regarding Time as stretching in the direction of his own, apparently straight, body line.

But Einstein's theory embraces a further supposition; one which, unfortunately, removes the subject to regions largely beyond the comprehension of the man-in-the-street. This 'Space-Time' extension is said to be not 'flat', but 'curved'.

The extensions of Time-extended objects are usually, in Relativity theory, called '*World lines*'; but they are sometimes referred to as '*Tracks*'. 'An individual', says Professor Eddington,[1] 'is a four-dimensional object of greatly elongated form; in ordinary language we say that he has considerable extension in time and insignificant extension in space. Practically he is represented by a line—his track through the world.' The addition of those last five words to an otherwise perfectly complete statement may seem to the reader something akin to 'hedging'—for how can the line be both the observer and the observer's path? But Eddington, a little farther on, is at pains to make his own view clear. The 'track' of the (presumably physical) observer is that observer '*himself*'. The italics are Eddington's own. And, again, lower down on the same page, he remarks: 'A natural body extends in time as well as in space, and is therefore four-dimensional.'

This seems plain enough. To any specific observer contemplating such a system of fixed, objective lines, the appearance of motion in the dimensions representing Space could be produced, as in Hinton's model, by the real movement, along the observer's 'track', of a field of observation apparently at right angles to the dimension representing Time. But to suggest anything of that kind would be to hint that this Time-travelling field of observation pertained to a *psychical* observer. For the physical observer is already defined as the 'track' travelled over.

Now, the Relativist has a very difficult case to present, and he certainly does not want to be handicapped with the burden of a psychical observer. On the other hand, he does not wish to appear to ignore the fact that we

[1] *Space, Time, and Gravitation*, p. 57.

observe events in succession. It is this quandary which drives him to a statement which appears to be intended as non-committal. The 'observer' is said to move along his 'track', and the reader is left to infer what he pleases from that.

Unfortunately, however, the reader has usually been allowed to infer that by 'observer' is meant a physical apparatus, inorganic or organic. So he can hardly be blamed for supposing that he is intended to understand that the 'track' is formed merely by the peculiar warpings of the Relativist's 'Space-Time', and that the physical elements of the observer's body move over the tracks, leaving these empty before and behind.

If, however, he were to assert that this is the teaching of Relativity, he would be told that a track which possessed reality in such a sense and to such an extent as to account for all the physical characteristics of an imagined three-dimensional object moving along it would be, in every one of its cross-sections, *physically indistinguishable* from the imagined object itself. Physically, the track would actually *be* the object extended in Time.

And that is the crux of the whole business. Anything that could properly be regarded as moving along the track would have to be something different from the fixed sections of the track itself.

Relativity theory, however, is quite clear upon one very important matter. *The Time dimension, for any given observer, is simply the dimension in which his own world-line happens to extend through the four-dimensional continuum.*

CHAPTER XIX

We are now, I think, justified in accepting two propositions:

1. That the brain contains memory traces of our past, attended-to experiences.

There seems to be no escaping this conclusion. Concussion does not destroy merely the ability to give verbal or other expression to the memories involved. The memories themselves are in some way affected, for the patient's mind appears to himself to be completely blank so far as these memories are concerned. And, since the physiological evidence is that such traces must in any case be formed, and must be destructible, we have no grounds on which to seek for any other explanation of the facts.

2. That Time may be treated as having length, divisible into years, days, minutes, etc.—a length in which each instant lies between two neighbouring instants—a length in which events are situated.

That is the classical conception. And its enunciation is equivalent to saying that Time is a fourth way in which length can be measured—a fourth dimension of Extension.

We are not, however, accepting Proposition 2 merely because it embodies the popular view. We do so because it follows logically from Proposition 1. For we have to recognize that a brain stimulation which is past, and a similar brain stimulation at a much later period, are not one and the same event, but two events separated by intervening events. We might have imagined that separation as being in some fourth-dimensional 'memory train'. But Proposition 1 rules out that idea. We start with the conception of memory as being merely the re-stimulation of an old brain trace. Hence we have to regard the separation of the two brain events as being in Time.[1]

Incidentally, this means that our Time length is not unoccupied; it contains physical configurations. This argument might have been useful, were it not that the reader is, I take it, already satisfied by the argument in Chapter XVII that the conception of Time as having length is utterly meaningless unless that length is regarded as occupied by such events. Moreover, if Time has length, the endurance of anything in Time must mean, as Wells pointed out, extension in that length.

* * *

We may note here that we need not trouble to debate the question as to whether the idea of Time as having length is an analytical device or the recognition of a 'reality'. Analytical devices are merely instruments for rendering manifest differences and relations which, without such assistance, would remain concealed. But unless these relations are already there, waiting to be brought to light, the analytical device can exhibit nothing new. It is true that such contrivances may describe phenomena in a language of their own—as the mercury column in a thermometer indicates degrees of temperature in terms of divisions of height, or as the mathematician represents variables in terms of x and y—but that does not affect the question. Whatever the analytical device exhibits must have its corresponding characteristics in the underlying reality; and that is all that need concern the man of science.

However, lest the reader should suspect that he is being manœuvred into a position he did not intend to adopt, it might be as well to point out this: All the practical, everyday questions

[1] See also the criticism of Bergsonism in the third section of the present chapter.

he asks himself regarding Time are questions embodying *the assumptions that Time has length, that states of the physical world are positioned along that length, and that he experiences these events in succession. The answers to those questions must, therefore, be given* in terms of those assumptions.

[It might, also, be advisable, at this point, to warn the reader against a conception which is in the nature

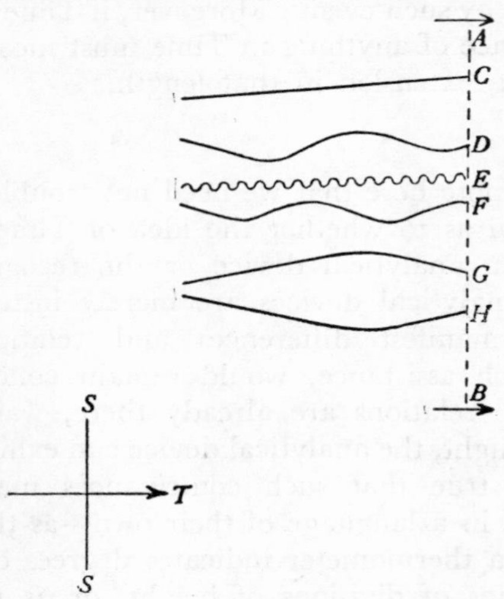

Fig. 2.

of a trap. 'Why,' it may be asked, 'do all these Time-dimensionalists, past and present, exhibit their physical "world-lines" as extending *ahead* of that "present moment" represented by *AB* in Fig. 1? Why should we not modify that diagram, and say that the world-lines are *growing* in Time, as shown in Fig. 2?'

The answer is that such a conception offends against the scientific law of the Economy of Hypothesis. That law forbids us to introduce, when considering a problem, more hypotheses than are strictly needed to cover the facts. For an unnecessary hypothesis is an *unwarranted* hypothesis.

Consider how the law applies in the present case. Fig. 3 represents the facts to be considered before we introduce the clarifying conception of Time's length. It represents a world of Space in which particles are moving about.

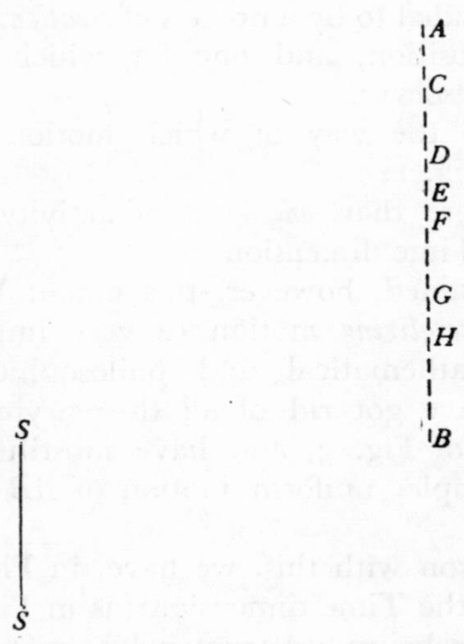

Fig. 3.

In this diagram we have—

1. Physical objects, *C*, *D*, *E*, *F*, *G*, and *H*.

2. Only *one* kind of activity—the motions of these objects up and down in the dimension representing Space. But these motions may be of varying velocity —a characteristic which we find it very difficult to comprehend or define. In fact, we had to wait until the days of Galileo and Newton before we could exhibit such varying velocities as determined by mathematical laws.

Now, introducing a Time dimension, we have, in Fig. 1 (see page 116):

The physical objects, one dimension larger than in Fig. 3.

In comparison with this, we have, in Fig. 2:

The physical objects, also one dimension larger than in Fig. 3. But, in addition—*unnecessary addition*—we have it that these extended objects must be conceived as being perpetually added to by a process of *creation*. This is a very strange proposition, and one for which we have no evidence whatsoever.

Turning to the way in which motion is exhibited, we have, in Fig. 1:

Still no more than *one* kind of activity—the motion of *AB* in the Time dimension.

We have gained, however, this much: We have succeeded in *generalizing* motion (a very important thing from the mathematical and philosophical point of view). We have got rid of all the varying, reciprocating motions of Fig. 3, and have substituted for these the single, simple, uniform motion of *AB* in the Time dimension.

In comparison with this, we have, in Fig. 2:

Activity in the Time dimension (as in Fig. 1); for the world-lines are being constructed by uniform growth in that dimension.

But we also still have activity in the Space dimension (as in Fig. 3); for the world-lines are being constructed by growth in that dimension as well as in the Time dimension.

Moreover, we still have the original complexity of motion of Fig. 3; for the growings in the Space dimension are of varying velocities.

Thus, while Fig. 1 involves the minimum of hypotheses necessary to cover the facts, and, incidentally, reduces motion to its simplest aspect, Fig. 2 introduces an additional, and so quite unnecessary, hypothesis—an hypothesis, moreover, which, instead of simplifying our idea of motion, adds further complexity thereto, and an hypothesis which is, in itself, of an extremely dubious character.

So our choice lies between Fig. 3 and Fig. 1, according as to whether we do or do not want to analyse the significance of Time.

The reader will, perhaps, forgive me if I conclude this chapter with a section addressed more particularly to students of Bergsonian philosophy.

* * *

[Fig. 2 seems to me to represent with absolute accuracy the conception of Time finally adopted by Professor Henri Bergson in his essay published the year following the appearance of Hinton's monograph. The date is of interest as showing that the fourth-dimensional theory of Time was well to the fore in those days.

Bergson begins by considering the supposed four dimensions of Extension—three of Space and one of 'duration'—and argues that the last-named is spurious. From this one is apt to assume, rightly or wrongly, that by 'duration' is meant Time, and that Bergson is attempting an analysis of Fig. 3 without employing the device of a Time dimension.

The moments of 'pure duration', he holds, are not external to one another, but are 'superposed', presumably as a printer might superpose pictures.

Presently, however, it becomes clear that 'pure duration' is not 'Time'.

'To sum up,' he says, 'every demand for explanation in regard to freedom comes back, without our suspecting it, to the following question: "Can time be adequately represented by space?" To which we [*i.e.*, Professor Bergson] answer: "Yes, if you are dealing with time flown; no, if you speak of time flowing." '[1]

And that, obviously, is what is represented in Fig. 2. 'Pure duration' thus seems to be identifiable with the man-in-the-street's 'present' and Hinton's moving 'narrow space and single moment'—the line *AB* in Fig. 2.

But Bergson sees that this acceptance of a Time dimension with moments which are external to one another is not enough. His 'pure duration' also has its moments, and these are *not* external to one another, but superposed.

He leaves us, thus, to contemplate two sets of moments—those which are superposed, and those in the 'past' part of a Time dimension.

Speaking entirely for myself, I should say that Bergson's superposed moments of 'pure duration' are his acknowledgments of the existence of that Time embracing Time which insists on obtruding itself into every attempt at temporal analysis. His growing 'past' takes Time to grow. But it would seem that Bergson, unwilling to recognize such a series of Times, and compelled by his earlier pages to grudge every inch of extension to any sort of Time whatsoever, has to take refuge in the 'superposition' idea.

Professor H. Wildon Carr (*vide* p. 114 of *The Philosophy of Change*) seems to exhibit Bergson's theory in a slightly different light, the element in Fig. 2 which grows as a train of past events being called 'Memory'. Remembering is, thus, a backward jumping of consciousness in a memory dimension. This theory, presumably, is what compels Bergson to devote so much time to a courageous, if rather forlorn, attack upon the accepted physiological view of memory. But Fig. 2 serves equally well to illustrate Wildon Carr's interpretation. We have merely to change the *T* in the dimension-indicator into an *M* standing for

[1] *Time and Freewill*, p. 221.

memory, and to label our moving line *AB* as *DD* standing for Bergson's 'pure duration'.

In either case the diagram stands condemned for the same reason as before: it introduces the totally unnecessary hypothesis of continual creation out of nothing, in addition to extension in a fourth dimension; and this at the cost of still further increasing, instead of simplifying, the complex character of variable motion.

Bergson's attitude as regards future events is emphatic. As in Fig. 2, they simply do not exist in any shape or form whatsoever. His argument for freewill is based upon that.]

PART V
SERIAL TIME

CHAPTER XX

A 'Series' is a collection of individually distinguishable items arranged, or considered as arranged, in a sequence determined by some sort of ascertainable law. The members of the series—the individually distinguishable items—are called its '*Terms*'.

The nature of the terms, when these are considered apart from their standings as members of the series, is of small consequence to the mathematician. The terms may be, let us say, peas in a pod, or the oscillations of a pendulum, or ridges and furrows in ploughland, or the stresses along a cantilever girder—it is all one to him. His interest is concentrated on the *relation* between the terms—the relation which links each term to the next and makes manifest the law that binds the whole into an ordered extension.

This characteristic relation between the terms may or may not affect the values of the terms themselves. Thus the essential significance of a pea is not, that I know, greatly affected by the fact that it lies in a row of similar peas. But each swing of the pendulum owes the extent of its movement to the previous swing. And the stresses at any place in the cantilever girder, due to an applied load at the girder end, depend for their magnitudes upon the particular relation connecting the series of stresses along the structure. (For instance, in the simple beam shape, the values of the forces acting upon the uprights and diagonals constitute series of equal terms; but the values of the forces acting upon the longitudinal members constitute series in arithmetical progression.)

In the *first* term of a series, the relation which links the terms is absent on one side; and this lop-sidedness may have a very practical significance. Thus the first swing of the pendulum has no previous swing to determine it: it must be started by an external agency. The first furrow in a ploughed stretch differs in section from all the others. And the forces acting on the end members of our cantilever girder are balanced at the outer ends, not by pushes and pulls in similar members, as elsewhere in the series, but by the externally applied end-load.

Now, we have seen that if Time passes or grows or accumulates or expends itself or does anything whatsoever except stand rigid and changeless before a Time-fixed observer, there must be another Time which times that activity of, or along, the first Time, and another Time which times that second Time, and so on in an apparent series to infinity. And we might suppose that every philosopher who found himself face to face with this conspicuous, unrelenting vista of Times behind Times would proceed, without a moment's delay, to an exhaustive and systematic examination of the character of the apparent series, in order to ascertain (*a*) what were the true serial elements in the case, and (*b*) whether the serialism were or were not the sort of thing that might prove of importance. For, of course, it might turn out to be an entirely negligible affair. But, to people who have devoted their lives to the search for a simple explanation of the universe, the idea that one of their approximate fundamentals—next door, indeed, to the sought-for nothingness—might prove to be of a serial character would be bound to appear a supposition to be avoided at almost any cost. Quite rightly, they would pause, and look round for some shorter path. Yet to a halt of that kind one is obliged to set a limit. To stand, for twenty-two centuries, staring at a perfectly open road is not necessarily at variance with the recognized traditions of philosophical procedure. But it would be a pity to risk having this estimable circumspection mistaken for commonplace somnolence.

CHAPTER XXI

Whether we embark upon the analysis of a serial time because of the logical compulsion, or whether we do so from motives of curiosity as to what sort of a country such an avenue would be likely to reveal, we must realize that, if we discover anything which is not already manifest in the ordinary, accepted first stage of the series, that thing will be something outside the purview of any philosophy which has been developed upon the basis of a uni-dimensional Time. That is to say, it will be something entirely strange to our present views of existence. We shall have, therefore, no right to halt and haver merely because we encounter novelty—novelty is what we are expecting to find. We must bear in mind, moreover, that serialism in Time is almost bound to signify serialism in other matters. In actual fact (the reader had best be warned of the worst) we shall find that it involves a *serial observer*.

In these circumstances the strictly proper course will be for us to get the analysis finished first—regardless of whether what is exhibited appears as fantastic or otherwise, so long as it follows logically from our premises—and *then* proceed to ascertain whether the results do or do not assimilate with the general body of our knowledge. And, as it happens, this is one of those cases where the adoption of a correct method is imperative; for it is not until the analysis is finished that the new conceptions begin to assume any sort of complete significance.

The reader, then, is advised to put all thoughts of meanings and implications entirely out of his mind until we come to the next chapter, and to regard the present analysis as a simple mental exercise of no more actual import than a cross-word puzzle. So that all he will need to do for the moment is to satisfy himself that the laws recited at the conclusion of this chapter are laws which have been properly deduced from our premises, and that they represent quite truly the relations between the terms of our series.

* * *

'From the windows of our railway carriage', says Professor Eddington, 'we see a cow glide past at fifty miles an hour, and remark that the creature is enjoying a rest.'

This is an illustration which pleases in more ways than one; and I regret to have to interrupt the reader's contemplation thereof in order to direct his attention to a picture painted in less enticing colours. But we have to get on.

We are still, then, seated in the same carriage; but this is now standing at a railway station. Looking from the windows on the side remote from the platform, we perceive another train at rest upon the rails. As we watch it a whistle blows, and we become aware that our train is beginning to pull out. Faster and faster it goes; the windows of the opposite train are running swiftly across the field of view; but . . . a doubt arises . . . we miss the accustomed vibration of our vehicle. We glance towards the platform windows, and discover, with something of a shock, that our carriage is still stationary. It is the other train which is moving.

Now, in the first of these two cases attention is fixed upon the visual phenomenon of the cow; this phenomenon moves across the 'field of presentation', and attention follows it. We judge that attention is directed to a point in the field of presentation corresponding to something which is fixed in external Space; and that, while attention is thus fixed, the field of presentation, *and the observer*, move in relation to that Space.

In the other instance, again, the visual phenomenon of a window pertaining to the opposite train moves across the field of presentation, and attention follows that phenomenon. Again we judge that attention is fixed and that the field—*with the observer*—is moving; but afterwards, in the light of other evidence, we reverse that judgement and say that the field and observer must have been fixed, and that attention must have moved.

In each case, then, the judgement may differ; but in each case the direct psychological *experience* is of the same general character. The phenomenon observed, whether this be the cow or the window of the opposite train, moves across the field of presentation—followed by the focus of attention—until it disappears at the edge of the field. *And in each case the field of presentation remains fixed with regard to the observer.*

Such a field of presentation, fixed with regard to the observer, and in which *observation, condensed to the shifting focus called 'attention'*, is assumed to be taking place, is bound to be the starting-point of our analysis. (All readings of instruments are perceived as appearances within that field.) It must be remembered, however, that the field contains phenomena other than visual; it embraces, in fact, every species of mental phenomenon which, whether attended to or not, is being presented for observation. It represents the observer's *outlook on Space*. And, according to the theory of psychoneural parallelism (see page 21), it occupies the same spatial position as daes that portion of the observer's cerebrum which is in the state of apparent activity associated with the production of observable psychical phenomena.

We shall represent this spatial position of field and cerebrum by *CD* in Fig. 4, the up-and-down dimension of the paper being regarded as Space. Temporal measurements are not yet shown.

Since the contents of *CD* are to be considered as in a state of apparent activity, they must be imagined as apparently moving up and down in the dimension representing Space. Moreover, the length of *CD* is uncertain; for larger or smaller portions of the cerebrum may be active at different instants. The diagram is to be looked upon, in fact, not only as a model, but as a *working* model. We indicate this by fitting two little arrowheads to the dimension-indicator at the bottom of the

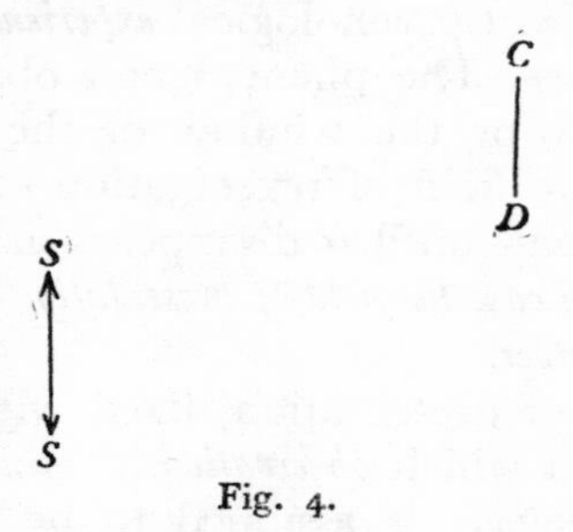

Fig. 4.

diagram, showing that motion in Space is supposed to be taking place.

(It must be remembered that, according to the more commonly accepted view of Space, *CD* itself may be moving as a whole in the Space dimension.)

Fig. 4 is our starting-point. It does not represent a 'term' in the Time series; for Time is not being indicated therein at all.

To the observer whose field of presentation occupies the spatial position *CD*, events are presented in succession. To him Time is apparent as an insistent characteristic of existence—a characteristic which, though real enough to be of immense personal importance, cannot be defined in terms of the three-dimensional limits of his spatial outlook. Phenomena in his field seem to move about, alter, and vanish. And these changes appear to 'take Time'. He endeavours to identify this 'Time taken' with a bit of Space moved over by some indicator such as a clock-hand; but fails because he cannot rid himself of the knowledge that the movement of the clock-hand is not measurable in terms of the clock-face alone. The hand 'takes Time' over its movement: it may traverse the clock-face quickly or slowly. Stopping the clock does not prevent other movements from 'taking Time'. He is aware of a growing store of memories; but is certain that this growth is also a process which 'takes Time'. Even when he sits in the dark and thinks, he is aware that such thinking is 'taking Time'. And when he recovers from an anæsthetic, he has evidence that Time has 'elapsed'.

He realizes that this 'Time' which is 'taken' is a measurable thing; that the measurement involved is of the simple, one-way kind called 'extension'; and that in this extension, the phenomena he observes persist for longer or shorter lengths. And, since we are in entire agreement with him, we will proceed to introduce this dimension of extension into our diagram as the side-to-side dimension of the paper.

The total process may be more easily followed if we divide it into two half-steps. The first of these consists merely in showing the physical elements in the cerebrum *CD* as having extension (*i.e.*, *endurance*) in Time. We begin by taking an instantaneous photograph of Fig. 4. To avoid trouble with the Relativists, we shall assume that we are standing side by side with the proprietor of *CD*. We may, thus, consider the positions which the moving elements exhibit in that photograph as their position at that particular instant of Time which both we and the owner of the cerebrum in question regard as the 'present' instant. This photograph is shown as *CD* in Fig. 5, the dotted prolongation of this line indicating this present instant. We show the 'past' and 'future' states of the moving elements of Fig. 4 as occupying fixed positions to, respectively, the left and right of *CD* in a Time dimension. These 'past', 'present', and 'future' states will together give us a band of wavy lines enduring (extending) in Time.

But although 'past' and 'future' states of the cerebral elements are shown as entities occupying fixed positions in the Time dimension, it is questionable whether we

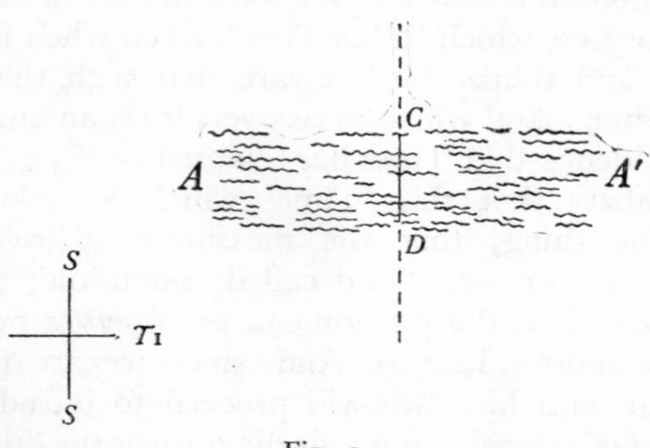

Fig. 5.

are treating the field of presentation in the same way. The fact that *CD* in Fig. 5 is a snapshot of the moving elements of Fig. 4 at an instant of Time which both we and the owner of the cerebral elements photographed regard as the 'present' instant, seems to suggest that *CD* is the only field of presentation in the whole extension.

Let us consider this question more closely. We have now accomplished the first of our two half-steps, and it will be seen that the result is to leave us with a very incomplete representation of the state of affairs which we started to analyse—the state exhibited in Fig. 4. The elements in that diagram were considered as *moving* up and down in the Space dimension, such motion being accompanied by changing psychological phenomena apparent to the owner of the pictured cerebrum. The diagram was to be regarded as a working model, exhibiting its states in *succession*. But there is no evidence of any appearance of change to any observers in Fig. 5. The lines which show the elements of Fig. 4 in their Time extension—the band *AA′*—are considered as being stationary in all dimensions. (For that reason we have

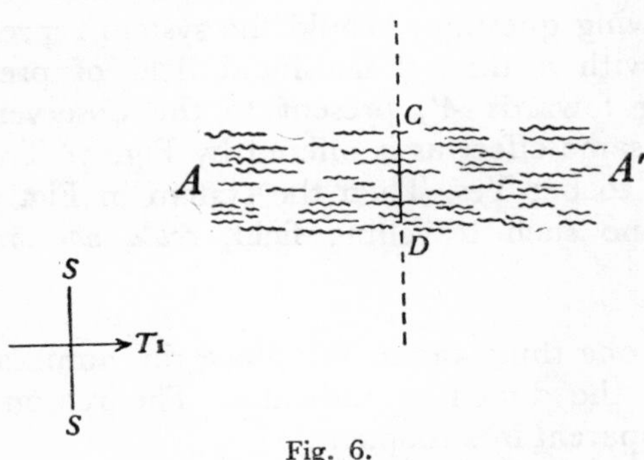

Fig. 6.

had to remove the arrow-heads from the little dimension-indicator.) And the cerebral states represented by the various cross-sections of that band are not being presented to any observer in succession. Either they are all being presented together, or else one only is being presented—the state at the 'present' instant, *CD*.

The second of our two half-steps consists in the re-introduction of these missing phenomena of motion. We do this in the obvious and, indeed, only possible way—the way to which the reader is now accustomed. We simply add an arrow-head to the *T* in the dimension-indicator, in order to show that *CD* is—as we had all along suspected—the only field of presentation in the diagram, and that this field is travelling along the Time dimension in the direction indicated by that arrow. We do this in Fig. 6.

[*Note to Third Edition.*

It is obvious that the whole demonstration of what we may call the 'regressive' character of Time hangs upon the validity of the foregoing argument. The reader who suspects the presence, in these two paragraphs, of some obscure confusion of thought, should ask himself the following question: Would the system represented in Fig. 6, with a three-dimensional field of presentation travelling towards *A′*, present to the observer of that field the same effect as is offered by Fig. 4? The answer is bound to be: Yes. Then the system in Fig. 5, where there is no such travelling field, *could not provide that effect.*]

We do one thing more. We place the numeral 1 after the *T* in the dimension-indicator. The reason for this will be apparent in a moment.

The first stage of our analysis is now complete, and it brings us to a merely revised edition of our starting-point. Our diagram is again a working model, and it no longer contradicts the statements we made regarding Fig. 4. The line *CD* is still, as we had originally stated, a field of presentation. Events are being presented in succession within that field. And the intersection points between that travelling field and the wavy lines are moving up and down within the field, providing for the observer effects of ordinary spatial motion.

As the field of presentation moves over the extended substratum, some of the phenomena presented in the field will appear as moving in relation to other phenomena in the field. For attention, focused upon the apparently moving phenomenon, has a fringe which covers enough of the immediately adjacent, comparatively non-moving phenomena to enable the difference to be perceived.

The result of this first stage leaves us, however, still dissatisfied. Analysing what was involved in our premises, we have arrived at conclusions which, so far as they go, are logically unescapable. The trouble is that they do not go far enough.

To begin with, we find ourselves confronted with a new object for consideration: to wit, a Time-travelling field of presentation.

Now, we cannot separate, in the Time dimension, that travelling field of presentation from an observer to whom its contents are being presented—contents provided by the cerebral elements in the substratum travelled over. Also, we are bound to regard this observer as three-dimensional. And, to avoid any possible confusion, we had better set forth exactly what that statement implies.

A Time dimension, for any observer, is a dimension in which *all* the events which he experiences appear to him to follow one another in a definite sequence—a dimension in which he (or his attention) does not move backwards so as to upset that order of successive experience. Those dimensions in which his attention can move to and fro appear to him, therefore, to be at right angles to that Time dimension. Whatever dimension, then, in our diagrams, actually determines, for the observer moving therein, that order of successive experience, is that observer's true Time dimension.

To the observer we are here considering, the dimension which thus determines the order of his successive experiences is the dimension moved over by the field. The to-and-fro movements of his attention are, therefore, confined to the three spatial dimensions at right angles to that Time. So he is an entity whose *capacity for such observation* is three-dimensional. And that is what we mean by calling him a three-dimensional *observer.*

Whether he has, or has not, in other capacities, extensions in other dimensions is immaterial to the arguments in this chapter. As an *observer* he is three-dimensional.

Clearly, then, the field *CD* must be regarded as the place where this observer, travelling in the fourth dimension, intersects with *AA'*.

And here we get a clear view of the nature of the Time regress. Enlarging our conception of the universe by introducing Time as an additional dimension compels us to introduce also *the observer of the universe formerly considered.* This does not, be it noted, involve the absurdity of putting him into the world which he himself observes and describes (we do not try to insert this observer in Fig. 4). The picture is a new and larger one. But we are accepting the picture which *he* drew (Fig. 4) as valid—so far as it goes.

It is to be noted, however, that there is nothing in all this which need alarm the materialist. It is abundantly clear that, when this observer, with his field, reaches the terminus of the cerebral substratum, he will find that the observable phenomena have come to an end. Nor is there anything yet to show that he has the smallest capacity for interference with the sequence of the cerebral states which he observes.

Now, our first stage has left us with a new Time problem to consider. For the observing entity, with its field *CD,* is travelling neither so slowly as to be stationary, nor so rapidly as to be in all places at once; and every condition between those two extremes must be describable in terms of Time taken per distance traversed. But the distance traversed is along our first-considered

Time dimension; so the Time which is taken must be a Time which is not shown anywhere in the diagram. Just as our first-considered Time is not indicated anywhere in Fig. 4. Hence we mark the *T* in Fig. 6 as *T* 1, to show that it is not the ultimate Time which times the movements, real or apparent, in those diagrams. That ultimate Time we may call Time 2.

* * *

In order to simplify our next diagrams, we shall now draw the band *AA′* as it would appear to an eye set level with the page and looking up that page from bottom to

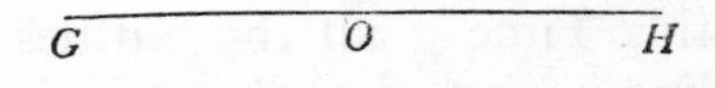

→ *T* 1

Fig. 7.

top. Seen thus, the band would appear as a single line; and this line is represented by *GH* in Fig. 7. The field *CD*—the place where our travelling observing entity intersects—is represented by the travelling point *O*. And each fixed point between *G* and *H* represents a single cerebral state, a spatial cross-section of the band *AA′*.

The Space dimension shown in Fig. 6 is here sticking out at right angles to the page. We shall have no room in the picture for other Space dimensions; but we may remember that they are supposed to be intersecting the diagram.

The view of affairs represented in Fig. 7 may be regarded as the first 'term' of our series. Time is exhibited and analysed therein, and it is shown that it is not ultimate Time.

Our business is now to exhibit the Time taken by the movement of *O* from left to right of Fig. 7 in exactly the same fashion as we exhibited the Time taken over the Space movements of the elements in Fig. 4.

The new dimension of Time will have to be at right angles to *GH*, just as our original dimension of Time had to be made at right angles to *CD* in Fig. 4. We shall, as already said, call this new dimension Time 2. In relation to this Time 2, Time 1 is, theoretically, akin to any of the three 'ordinary' dimensions of Space. Instead of a four-dimensional world in which the fourth dimension is Time, we have now a five-dimensional world in which the fifth dimension plays that insecure rôle.

In this Time 2 all the entities in *GH*, including the travelling entity at *O*, have endurance. That is to say, they remain in existence while you watch *O* travelling. These endurances will have to be shown as extensions in the Time 2 dimension.

We begin, as before, by taking our instantaneous photograph of our new working model. This photograph is taken at what is, to us, the 'present movement' of ultimate Time—the Time which times the movement of *O* along *GH*—that is to say, Time 2. It represents the condition of Fig. 7 at that 'present moment'. We exhibit this photograph as *GH* in Fig. 8, the line *pp′* indicating the 'present moment' in question.

Next, we have to show the 'past' and 'future' (in this Time 2 dimension) conditions of the fixed cerebral states represented by the fixed points in *GH* as, respectively, below and above their 'present' condition in *GH*. Since these states do not change their position either in Space or in Time 1, their endurances in Time 2 must be shown as extensions *straight* up Time 2. They thus become, in Fig. 8, vertical lines extending up and down the page *with no limit either way* that we are, as yet, able to assign.

But we need treat only a few selected points in this fashion.[1]

We have now another entity to consider—the three-dimensional observing entity which intersects at the three-dimensional field *O*. In the 'present' condition of Fig. 7 (*GH* in Fig. 8) the point of this intersection is at the middle of the line. Since, however, this point is, in Fig. 7, travelling along Time 1, its positions in the 'past' conditions of that diagram must be shown more towards the *G'G"* side of Fig. 8, and its positions in the 'future' conditions must be shown more towards the *H'H"* side. Linking up these various points of intersection, we get a diagonal line like *O'O"*. This line will represent the endurance (temporal extension) of the intersecting entity.

Here we have to ask ourselves again the same question that we asked in stage 1. We have shown the 'past' and 'future' states of all the entities in our working model (Fig. 7)—including the intersecting entity at *O*—as extensions of those entities, occupying fixed positions in the 'past' and 'future' parts of the Time 2 dimension. But have we treated our original three-dimensional *field of presentation* in that fashion? We have shown its present state; but have we shown its past and future conditions?

The answer is: No. Our diagram shows nothing beyond the *endurances*—the fifth-dimensional *lengths*—of the entities considered. There is, in that figure, no three-dimensional observer possessing a field in which there is an effect of change. *O'O"*, who has four dimensions of magnitude, intersects with all the cerebral states in the substratum, but that does not provide this four-dimensional creature with the requisite single, unique three-dimensional field with changing contents. Yet, our diagram has to show, as did Figs. 4 and 6, that such a

[1] To treat all the points in *GH* thus would turn the new figure into a surface black all over.

field, with such an effect therein, is presented to the ultimate observer—the owner of the cerebrum.

We have not lost that field. It is still where we placed it, at the intersection of *O′O″* with *GH*. But the figure has failed to show that this single, unique field is *moving*. How are we to make good this deficiency in our picture?

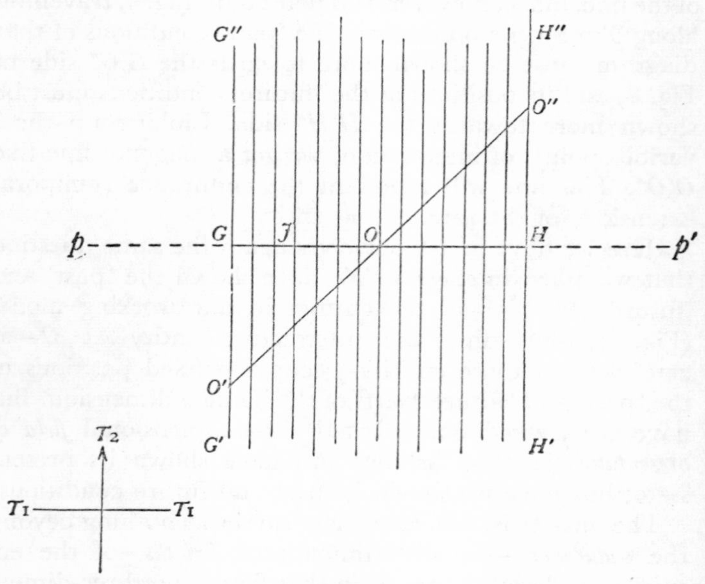

Fig. 8.

The field has to be regarded as travelling, and as travelling in Time 1. Since, while so doing, it must remain in *O′O″*, it must be considered as travelling up that diagonal; that is to say, as travelling up Time 2. Which means that, for our *ultimate* observer to observe the contents of the instants of Time 1 in succession it is necessary that he shall observe the contents of the instants of Time 2 in succession. He must have a field of presentation travelling up whatever is ultimate Time—in this case, Time 2.

By analogy with stage 1 we should expect that the *whole* of *GH* in Fig. 8 (the instantaneous photograph of Fig. 7 at a moment of Time 2 which appears to us as 'present') would turn out to be this field of presentation travelling up Time 2—a field the existence of which could not become evident until Fig. 7 had been expanded in Time 2. Just as, in stage 1, the existence of a Time-travelling field *CD* within the active Fig. 4 could not become evident until Fig. 4 had been expanded in Time 1.

It will be remembered, however, that the first term of a series may differ in some respects from all the remainder. Consequently it might be wiser not to trust to analogy here, but to continue to establish the characteristics of our second term by direct analysis of what is involved in the fact of succession in experience.

O, then, is travelling up *O′O″*. *But the only thing which marks off O as a definite point in O′O″ is the line GH.* This line, therefore, must be travelling up Time 2. *GH*, however, represents the condition of Fig. 7 at what we are considering to be the 'present moment' in Time 2. Hence this 'present moment' in Time 2 is travelling up Time 2.

It is advisable to remember here that, just as Time 2 is true Time in this stage, so is the Time 2 travelling 'present moment' the true travelling 'present moment'. Our old, Time 1, travelling 'present moment' has become merely an intersection point between the true travelling 'present moment' in Time 2 and a fixed diagonal in the diagram. It does not exist in its own right, but is determined by the Time 2 'present moment'. The point *O* is *determined* by *pp′*. To put it in scientific language, our Times are arranged in *series*, not in *parallel*.

Now, the points in *O′O″* are being consciously and *successively* observed from *O′* to *O″* by whatever is the ultimate observer. And we have just seen that the *only* thing which determines the order of succession in which these points are being observed is the travelling 'present moment' in *Time* 2. So the *ultimate* observer of the changing point in $O'O''$ is an observer for whom Time 2 plays the part of the real and only Time. Time 2 is the Time which determines the sequence of his experiences. This means that the Time 2 direction is the direction of travel of his field of presentation. And Time 1 is at right angles to what is, for him, the real and only determinative Time. Time 1 is, therefore, in relation to him, akin to a dimension of 'ordinary' Space. In other words, just as, in stage 1, the ultimate observer exhibited himself as a three-dimensional being in a three-dimensional world, so, in the more elaborate view afforded by stage 2, the ultimate observer exhibits himself as the four-dimensional observer in the four-dimensional world marked off by pp'. This four-dimensional observer must have a four-dimensional field of presentation lying in, and travelling in the same direction as, pp'. Clearly, then, he is not the entity $O'O''$.

But the discovery of new elements in our growing diagram does not entitle us to repudiate any previous supposition upon which that diagram has been erected. The argument for the existence of this field of presentation number 2 is *based* upon the hypothesis that there is a point O travelling in $O'O''$. And we may not now deny that $O'O''$ is, at O, a three-dimensional *observer*. For it is only because we acknowledge, in stage 1, the presence of such a three-dimensional observer at that point in GH that we were enabled, later, to insert the line $O'O''$ in the diagram. And so it goes from the beginning of the analysis. Nothing that has been previously ascertained and identified may be ignored later on. All that we may do is to discover new elements as our diagram grows more elaborate.

Hence, that three-dimensional section of *O′O″* which happens to be at *O* turns out to be an entity in the four-dimensional field of the so-far ultimate observer. We shall refer to this section of *O′O″* as 'Observer 1'. The four-dimensional observer may be called 'Observer 2'. And it is clear enough now that our analysis of the state of affairs with which we started is going to bring to light a whole series of such observers. The fact that observer 2 is *travelling* in the fifth dimension (moving up the vertical dimension in Fig. 8) means that that dimension cannot represent real, absolute Time, and will compel us to consider the Time which times the motion in question. When we introduce that Time as a sixth dimension, the earlier arguments will repeat themselves; and we shall unearth an observer 3 who will play, at that stage, the part of 'ultimate observer'. And there can be no end to that process.

Our immediate task, however, is to discover precisely how the observational activities of observers 1 and 2 are interrelated. That should not be difficult.

We began with the knowledge that the ultimate observer is affected by successive three-dimensional states of the substratum presented to him in the three-dimensional field *CD* of Fig. 4. That field has turned out to be the place in Fig. 8 where the four-dimensional field of observer 2, travelling upward, intersects the entity *O′O″*. And observer 2 is, so far, that ultimate observer who is being affected.

We have discovered, however, that each section of *O′O″* is being affected by one of the substratum states in question.

Now, the development of the series of observers places observer 1 (the section of *O′O″* which is at *O*) *between* observer 2 and the substratum section at *O* which is, somehow, affecting that observer 2. So that the process by which that particular state affects observer 2 is as follows. A certain feature in that state causes a corresponding modification in the intervening section of *O′O″*. It is this reproduced feature which affects observer 2.

But that raises the following difficulty. Observer 2 is a four-dimensional creature, and the section of *O′O″* which intervenes between him and the substratum is only three-dimensional. His field of observation must extend, therefore, in the fourth dimension beyond the place where *O′O″* crosses that field. In those outer parts of observer 2's field there are many other three-dimensional sections of the substratum containing the kind of feature which, reproduced in the intervening entity, is affecting observer 2. Since observer 2 is susceptible to features of that kind, what is there to prevent him from being affected by these other three-dimensional sections of the substratum as well as by the section of *O′O″* which lies in his field?

Nothing, that I can see. So, pending the discovery of some obstacle, we must assume that observer 2 is affected by the substratum adjacent to the section of *O′O″*. *But this collection of adjacent sections does not affect him in the same way that he is affected by the three-dimensional section of O′O″.* The bit of the substratum beside *O′O″* is a four-dimensional strip presented as a *whole* to a four-dimensional observer—it has, to him, no distinguishable three-dimensional sections. The function of observer 1 (*i.e.*, the function of the only purely three-dimensional entity within the field) is to abstract from the substratum an aspect thereof with which, otherwise, observer 2 could never become acquainted.

How far, now, can we say that observer 2's field (let us call it field 2) extends along *pp′*?

The answer is quite simple. Glance again at Fig. 8. This field 2 is, as we ascertained a moment ago, travelling in the Time 2 direction, *i.e.*, vertically up the diagram. As it does so, observer 2 observes in succession the various points in *O′O″ plus* a marginal bit of the substratum. We have shown the travelling field 2 as having just reached *GH*, with field 1 (*i.e.*, *O*) in the middle of that line. But, when the travelling field 2 was near the bottom of the figure, field 1 was at *O′*, and observer 2 was observing observer 1 *plus* a bit of the substratum at that point. And, when field 2 is nearly at the top of the diagram, field 1 will be at *O″*, and observer 2 will be observing at that place. Observer 2, therefore, must be able to observe each of the states in the substratum from left to right. So, since his field is moving vertically up the figure, that field must extend at least from *G* to *H*.

If, then, *G′G″* represents that state of the cerebrum where it first (in Time 1) becomes sufficiently developed to allow the ultimate observer to perceive psychological effects, and if *H′H″* represents the place where (in Time 1) that cerebrum ceases its useful activity and disintegrates, we may say that observer 2 can observe the *whole* of his ordinary, waking, Time 1 life, from birth to death, but that, for some reason to be determined, he allows his *attention* to follow observer 1 in that individual's journey from left to right (from birth to death) along field 2.

We shall need a name to distinguish *O′O″*, as a whole, from that section thereof which happens to lie within observer 2's travelling field and which is being employed by that individual as a source of information regarding the substratum which is affecting it. The other parts of *O′O″*—behind and ahead of *GH*—may also be affected by the substratum and may have served, or be about to serve, as instruments for observer 2; but, at that instant of absolute Time which we are examining, they are outside that individual's field and he is making no use of them. In the earlier editions of this book, *O′O″*, as a whole, was given the name of 'Reagent 1'. A reagent is a substance employable as a detector; and, although the word 'instrument' might be more suitable in this case, I have adhered to the older nomenclature to avoid confusing earlier readers. The section of this reagent or instrument which is conveying information to the ultimate observer is styled, as said already, observer 1.

We have arrived now at the conclusion that *GH* is, like *CD* in Fig. 6, a field of presentation. And, like those stage 1 fields, it stretches, athwart the Time dimension, from edge to edge of the cerebral substratum. Since this characteristic holds good in two terms of the series, we may regard it as a repetitive relation which will appear in every term.

We conclude stage 2, then, by fitting an arrowhead to Time 2 in the dimension-indicator of Fig. 8, in order to show that *GH* is a field of presentation moving up Time 2. The motion of field 1 along Time 1 is now recovered. For, as *GH* moves up the diagram, the point *O*, where *GH* intersects with *O′O″*, moves along *GH* towards *H*, thus coming upon the cerebral states one after another in succession from left to right.

Our diagram—which represents the second term of the series—is once again a working model. And it does not contradict the information previously provided by Fig. 7. In that figure, *O* was a point of intersection travelling along *GH*. Our more elaborate diagram confirms that statement, and merely supplies the additional information that the *travelling* of the intersection point is due to the Time 2 travelling of the *GH* which is intersected; that *GH* proving to be a field of presentation concealed in the over-compressed view afforded by Fig. 7. We still have at *O* our original three-dimensional observer moving along Time 1, but he proves to be merely a section of his own temporal extension above and below in the form of the diagonal reagent.

It is to be noted that the travelling observer at *GH* must be, in his turn, a line where an entity, reagent 2, intersects with the plane figure *G′G″H″H′*. Also that ultimate Time—the Time which times the movement of *GH* up the plane, and of *O* along *GH*—is not Time 2, but *Time* 3.

* * *

[We may, conveniently, carry the analysis one stage further; but we need not trouble to repeat the arguments.

We shall discover, of course, that the Time and the field and the observer, which, in stage 2, we considered as being ultimate, were not ultimate at all; and we shall come upon a larger-dimensioned lot of ultimates which, in their turn, will only retain that status until the next stage is reached. And so on to infinity.

In Fig. 9 we exhibit three dimensions of Time as the three dimensions of a solid figure seen in perspective. We have to draw imaginary boundaries to this figure in order to make the perspective clear; but *actually, there are no such boundaries at the top or the bottom or the back or the front. The figure has fixed sides* (*representing birth and death in Time* 1), *but its extensions in the Time* 2 *and Time* 3 *dimensions have no limits.*

Time 3 is shown as the vertical dimension of the block. In relation to this Time the dimensions we call Time 1 and Time 2 are akin to dimensions of Space.

The middle horizontal plane-section of this block-figure, the plane *G′G″H″H′*, is our instantaneous photograph of Fig. 8, shown in perspective. The endurances, in the new dimension of Time, of the cerebral states represented by the Time 2 extended lines in Fig 8. should be shown by extending these lines in the Time 3 dimension so that they form vertical planes arranged like pieces of toast in a rack. But to fill these in would overcrowd the diagram. Our first reagent, *O′O″*, will endure (extend) in Time 3 as a plane dividing the block diagonally; that is to say, the plane *ABCD*.

In the 'present' condition of Fig. 8 (shown in the middle of the block), the field of presentation *GH*—which, be it remembered, must be marked out by the intersection of some observing entity with the plane of the figure—is at the middle of the plane. In the 'past'

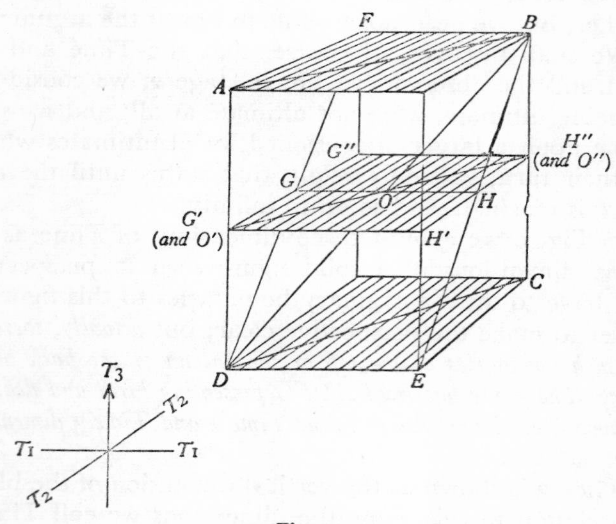

Fig. 9.

condition of Fig. 8 (the plane at the bottom of the block) this field—this line of intersection—is at *DE*. In the 'future' condition of Fig. 8 (at the top of the block) this field is at *FB*. The intersecting entity, reagent number 2, lies, therefore, along the sloping plane *DFBE*, which plane represents its endurance.

The intersection of this plane with the plane *ABCD* is the line *DB*. The new travelling field of presentation (field 3) is the plane *G′G″H″H′*. As this field 3 plane travels up the block, its line of intersection with the sloping plane *DFBE* (the line *GH*) moves over the travelling field 3 plane towards *G″H″*. That is to say, field 2 moves along Time 2. The point *O* (where the three planes *ABCD*, *DFBE*, and *G′G″H″H′* intersect) moves, meanwhile, along the travelling line *GH* towards *H*. That is to say, field 1 moves along Time 1.[1]]

* * *

The analysis will continue, evidently, in the same fashion for as far as you care to follow it.[2] There we shall have a single multi-dimensional field of presentation in absolute motion, travelling over a fixed substratum of objective elements extended in all the dimensions of Time. The motion of this ultimate field causes the motion of an infinite number of places of intersection between that field and the fixed elements, these places of intersection constituting fewer-dimensional fields of presentation. There, again, we shall have a Time which serves to time all movements of or in the various fields of presentation. This Time will be '*Absolute Time*', with an absolute past, present, and future. The present moment of this absolute Time must contain all the moments, 'past', 'present', and 'future', of all the subordinate dimensions of Time.

It will be noticed that we can never show the path which *O really* follows. In Fig. 8 this path appears as *O′O″*,

[1] It will be remembered that the figure is a diagrammatic representation of *serial* relations, and that one cannot, in considering movements within the block, overlook the system on which that block has been constructed. One cannot, for example, consider the point *O* as moving up *DB*, without, at the same time, recognizing the conditions of that movement, *viz.*, that field 3 is travelling in Time 3, and field 2 in Time 2.

[2] See remarks at the end of the last Appendix.

but in Fig. 9 it appears as *DB*. We have to show it differently with each introduction of another dimension of Time. But it will be seen that, to the observer of each specific moving field in any diagram that can be drawn, *O*'s path will appear to lie *within his field.* (For example, to the observer of the field *GH* in Fig. 9, *O* appears as moving from *G* to *H*.)

The nature of the series is now beginning to become apparent. It is akin to the 'Chinese boxes' type—the type where every term is contained in a similar but larger (in this case *dimensionally* larger) term.

Its laws may easily be ascertained. As the first we have—

1. *Every Time-travelling field of presentation is contained within a field one dimension larger, travelling in another dimension of Time, the larger field covering events which are 'past' and 'future', as well as 'present', to the smaller field.*

The second law brings in the serial observer.

We have seen that the contents of the instants of Time 1 can only be presented to a higher-order observer in succession on condition that the contents of the instants of Time 2 are being likewise successively presented, and so with the contents of the instants of all other Times in the series that you choose to consider. This higher-order observer is, therefore, the observer of the field of presentation travelling up the dimension of Time at that place in the series. As the observer of that field, he is the observer of all the lesser and contained travelling fields.

('Higher-order' does not mean an observer who is in any way *remote*, in either Time or Space. The observer in question is merely your ordinary everyday self, 'here' and 'now'.)

So for our second law we have—

2. *The serialism of the fields of presentation involves the existence of a serial observer. In this respect every time-travelling field*

is the field apparent to a similarly travelling and similarly dimensioned observer. Observation by any such observer is observation by all observers pertaining to the dimensionally larger fields.

Hence, since 'attention' is only a name for concentrated observation, the attention of the observer pertaining to any field must be referable to the attentions of the observers pertaining to the dimensionally larger fields. But the focus of attention (the area covered by observation of a given degree of concentration) must have, in each case, the same number of dimensions as have the observer and his field. In field 1 it is three-dimensional; in field 2 it is four-dimensional; and so on.

Consequently we have, as our third law—

3. *The focus of attention in any field has the same number of dimensions as has that field, and is a dimensional centre of the foci of attention in all the higher fields.*

And now let us see whether there is anything to be made of it all.

CHAPTER XXII

Our analysis has ascertained the nature of the temporal machinery which is bound to exist if we observe events in succession.[1] The question which has now to be answered is whether an inspection of that machinery will enable us to account for anything else. And the reply is in the affirmative.

How would you define rationally a '*self-conscious*' observer—define him so as to distinguish him from a non-self-conscious recorder such as a camera? You would begin, I imagine, by enunciating the truism that the individual in question must be aware that something which he calls 'himself' is observing. Putting this into other words, the assertion is that this 'self' and its observations are observed by the self-conscious person. But it is essential that he should observe his objective entity as something pertaining to *him*—he must be able to say: This is *my*-'self'. And that means that he must be aware of *a 'self' owning the 'self' first considered.* Recognition of this second 'self' involves, for similar reasons, knowledge of a third 'self'—and so on *ad infinitum.*

It is difficult to see how such a serial observer can exist anywhere in the three dimensions of Space alone, but the analysis in our last chapter has shown that he can—and does—exist very nicely in the multitudinous dimensions of Time. Reagent 1, reacting to the cerebral substratum at *O*, is, there, a presentation in the travelling field of observer 2—is a sectional feature which, in its state of reaction, is being observed. Similarly, reagent 2 is, at *GH*

[1] Note the distinction drawn here between events in a system observed, and observational events. The entire 'regress' of Time depends upon the fact that these two classes of events cannot be allotted positions in a single dimension.

(wherein it is observing at *O*), a presentation in the travelling field of observer 3 (*vide* Fig. 9, page 156). And so on for as far as you care to go.

But let us see if the analysis has yielded anything else.

Well, there is . . . but this is a lapse into pure psychology, and must be regarded as such. Psychologists are always seeking for an explanation of how it is that we are *aware* of the passage of Time—aware, that is, not merely of motion or of change, but of the fact that motion and change involve Time transit. That Time should be a length travelled over is, all said and done, a rather elaborate conception; yet that this is the way we do habitually think of Time is agreed to by everyone, both educated and—which is much more curious—uneducated. The child instantly understands its nurse's lumbering attempts at explanation. It scarcely needs to be told that 'yesterday' has 'passed by' and that 'to-morrow' is 'coming'. How does it, how did we, arrive at this remarkable piece of knowledge?

A theory often hazarded is that attention is never really confined to a mathematical instant. It covers a slightly larger period. That is to say, it has a small extension in the Time dimension.

Now, this small extension is actually given us by Law 3 of the series. The law asserts that the focus of attention in field 1 is the dimensional centre of the foci of attention in all the higher fields. That means that the focus in field 1 is surrounded by a fringe which, however narrow it may be, is being subjected to attention by observer 2. That means, again, that observer 2, whose attention is surrounding and following observer 1's attention in field 1, must perceive observer 1's apparent movement in relation to those stationary (cerebral) presentations in field 2 which are covered by his own dimensionally larger focus.

The process is precisely similar to that by which observer 1 perceives objects travelling across his own three-dimensional field. Hence observer 2 not only observes what observer 1 is observing, but perceives that individual as travelling from 'past' to 'future' in Time 1.

(Philosophers will note that 'succession in experience' is thus bound to involve the 'experience of succession'.)

In connection with this overlapping of the focus of observer 1 by that of observer 2, there is another point which may possess some measure of significance. Any focus of attention travelling along Time 1 will come upon irregularities in the substratum—irregularities which we represented by the waviness of the substratum lines in Fig. 6. In their relation to field 1 and its focus, these irregularities, whether observed or not, are the movements of physical elements in three-dimensional Space. But the slightly wider, overlapping focus of observer 2 may quite well cover a Time 1 length containing a considerable number of these irregularities, which would thus be presented as Time 1 *pattern* in the part of the substratum covered by that focus. This means that observer 2, following field 1 with his attention, should be capable of directly perceiving in the objective universe characteristics beyond those which present themselves as the spatial groupings and spatial movements of enduring particles. Physical *frequency* would be presented as pattern—a frequency would appear as something concrete. This may ultimately prove to have some formal connection with the observer's interpretations of frequency as sensation. But we should, probably, be exhibiting the matter in its most significant aspect if we said that what the ultimate observer should thus be able to observe directly is a highly important and very remarkable characteristic known to physical science as '*Action*'. But of this more anon.

Can we gather anything else?

Yes, we have, at last, the explanation of our dream 'effect'.

Law 3 asserts that the focus of attention in any lower field is surrounded by the foci of attention in all higher fields. Thus, in waking moments, the attention of observer 2 is not ranging to and fro over the limits of field 2, but is following the focus of observer 1 in field 1 moving

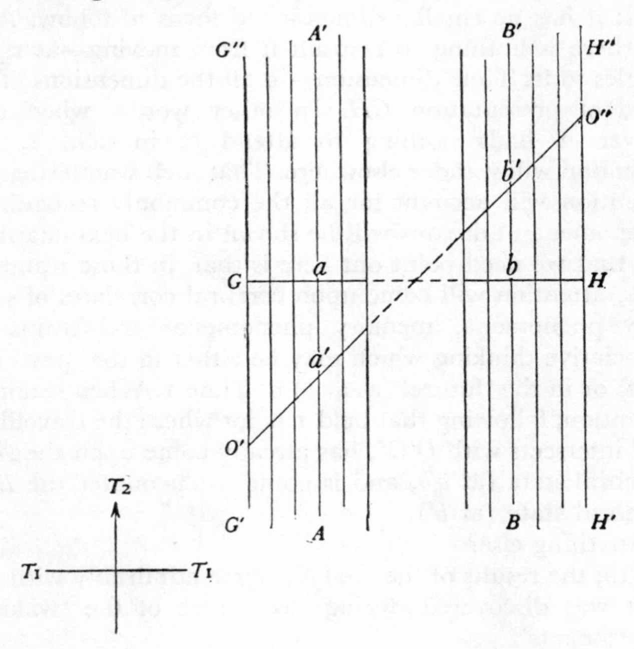

Fig. 10.

laterally across field 2. But what if there is no focus of attention in field 1? What if field 1 becomes, as in deep sleep, a blank, owing to the passivity of the cerebrum? Such a situation is exhibited in Fig. 10.

The gap running up the middle of the diagram indicates the absence of all such cerebral states as are associated with the production of psychical phenomena. So, at the moment (in absolute Time) when field 2, *GH*, moving up Time 2, is in the position shown, there is nothing in field 1 (the intersection point of *GH* and *O′O″*) for observer 1 to attend to. The focus of attention of observer 2 has thus become the *first term* of the series of concentric foci: it has no smaller-dimensioned focus to follow. And so there is nothing to restrain it from moving—at right angles to its Time dimension—in all the dimensions of its field of presentation *GH*. In other words, when observer 2 finds nothing to attend to in field 1, his attention will wander elsewhere. That such wanderings of attention will account for all the commonly recognized phenomena of dreams will be shown in the next chapter. All that we need point out here is that, in those wanderings, attention will come upon cerebral correlates of sensory phenomena, memory phenomena, and trains of associative thinking which may be either in the 'past' (as at *a*) or in the 'future' (as at *b*) of Time 1. When waking, attention, following that field 1 point where the travelling *GH* intersects with *O′O″*, has already come upon the *AA′* cerebral state (at *a′*), and is going to encounter the *BB′* cerebral state (at *b′*).

Anything else?

Yes; the results of the analysis agree admirably with all that was discovered during the course of the 'waking experiments'.

That analysis has sharply distinguished *presentation*, referable to the original cerebral states, from *observation* (which includes attention), referable to the observer. So it is not surprising that it has brought to light no law which *compels* the observer last considered to direct his attention to any particular phenomenon in any particular field. That such attention is, as a matter of plain fact, habitually directed during waking moments to phenomena in field 1 is obvious enough; but theory leaves us with habit as the only compulsion in the matter. And practice bears this out. In the waking experiments, as the reader will remember, attention, so long as it was allowed to follow an easy, swift train of associated images, came upon nothing but images of the past. The reason now seems fairly clear. That the train of associated images came into observation swiftly and easily showed that the attention of the ultimate observer was travelling according to habit. But habit keeps it in field 1, and in that field all images relate to the past. Nevertheless, the habit was no *law*. It could be overcome. By determinedly refusing to attend to these readily proffered images, attention in field 1 could be completely discontinued. And, in the rare instants when this was successfully effected, attention in field 2 was free, as in dreams, to slip away along associational tracks extending elsewhere than in the Time 1 'present moment'.

Confining ourselves, in this chapter, to the simplest things deducible from the analysis, we have one more point to note.

It is abundantly clear that our serial observer is going to have considerable difficulty in disengaging himself from the trammels of self-conscious existence. In fact, one cannot see how he is going to manage it at all.

The substratum which provides the ultimate contents of his serial field of presentation is merely the extension (endurance), in many dimensions of Time, of the primary extension in Time 1. That Time 1 extension has a beginning and an end, and these two boundaries are taken into account and appear everywhere in the extensions in the other dimensions of Time. But the fields which travel over the extensions in the second and 'higher' dimensions of Time *do not, in any term, move from or towards those two boundaries; they travel straight up between them.* The only field which runs out of the multidimensional figure is field 1. Death—that is to say, the arrival of a travelling field at a boundary—is, thus, not a serial element. It is, like sleep-gaps and the various Time irregularities in the substratum, one of those *solely first-term characteristics*, which—as we saw earlier—must exist in any series which has a beginning.

There may be, of course, arbitrary terminations to the extensions of the substratum in the other dimensions of Time—some deity may cut them off—but the analysis indicates that, failing such interference, the substratum persists to infinity in all Time dimensions save the first. For it does not exhibit in those other dimensions the characteristics which, in Time 1, indicate a possible splitting apart of the Time-extended lines at a place farther on in the stretch.

So observer 1 seems to be the only observer who ceases to observe.

* * *

The reader will note, I hope, that the foregoing tenets of Serialism have *not* been deduced from the empirical evidence supplied by our dream effect, but have been obtained by a direct analysis of what must, logically, be the nature of any universe in which Time has length and in which states of the external world are observed in succession.

The case for the dream effect is, therefore, a double one —logical and empirical. The procedure in the book might, indeed, have been entirely reversed. We might have begun by analysing what was involved in the fact that we experience events in succession. At the conclusion of that analysis we should have noticed—as a very trivial corollary to the disclosures of real importance—the probability of the dream effect. And we might then have described the experiments undertaken to test the validity of this last conclusion. That would have been the usual fashion of a scientific report.

But the circumstances in this case are unique. It is obvious that, although the 'higher-order observer' is nothing more magnificent or more transcendental than one's own highly ignorant self, he is beginning to look perilously like a full-fledged '*animus*'. Now, it has been pointed out, in Part I, that belief in the animus must have originated in the study of dreams. Savages and men of poor education, remembering their dreams, could have come to no other conclusion than that, in dreams, they were in a field of existence entirely different from that of ordinary waking life. That belief has been supposed to be childish and absurd. If it were really so, then the case for the animus would have to be regarded as tainted at its source.

I have thought it correct procedure, therefore, to begin by putting the savage before the court, and by showing, empirically, that his dreams did, in fact, occasionally provide him and his 'seers' and his 'prophets' with ample grounds for the belief that the dream field was something quite other than the waking field, and that his ultimate self enjoyed a degree of temporal freedom denied to the waking individual.

The proofs advanced in the present fourth part of the book can then be dealt with on their merits.

CHAPTER XXIII

Since all observation is the observation of the higher-order observer, all successive, automatic experience of the cerebral states situated along Time 1 is the thinking of that not always very clear-minded individual. But is this inspection of field 1 the only sort of thinking he achieves? And is what is presented in that field always so purely automatic as we have assumed throughout the previous analysis?

This higher-order observer (who, be it remembered, is merely your ordinary everyday self) observes in field 2 (*GH* in Fig. 11) an image *b* pertaining to a brain-state *bb′*, which state (vertical line) has not yet been reached by the intersection point between *GH* and *O′O″*. In other words, you dream of a future event, and this event is experienced, waking, a day or two later, when field 2 has moved to *G″H″*. On the morning following the dream—that is, when field 2 has moved only to *G′H′*—you, for reasons good or bad, note down on a piece of paper what you dreamed.

The memory trace of that dream-experience of *bb′* is, clearly, not in the brain-state at *cc′*, where field 1, *O*, is situated at the moment of writing down the dream. Therefore—to be extremely logical—it must be somewhere else.

The act of writing down the dream from that memory is thus a plain interference with the *automatic* sequence of cerebral events in Time 1. (How far this interference will affect our diagrams is a matter which will be dealt with in the next chapter.) Also, the total process of reasoning which selects certain details of that dream-memory (which is not in field 1) as being of importance to your intellectual investigation cannot be merely an inspection of brain-states in field 1.

We are therefore obliged to allow you the use of memory traces and intellectual equipments which are *additional* to those observable in field 1.

What can we discover about these?

Consider what happens when you fall asleep.

Your focus of attention becomes a four-dimensional focus confronted with four-dimensional presentations—

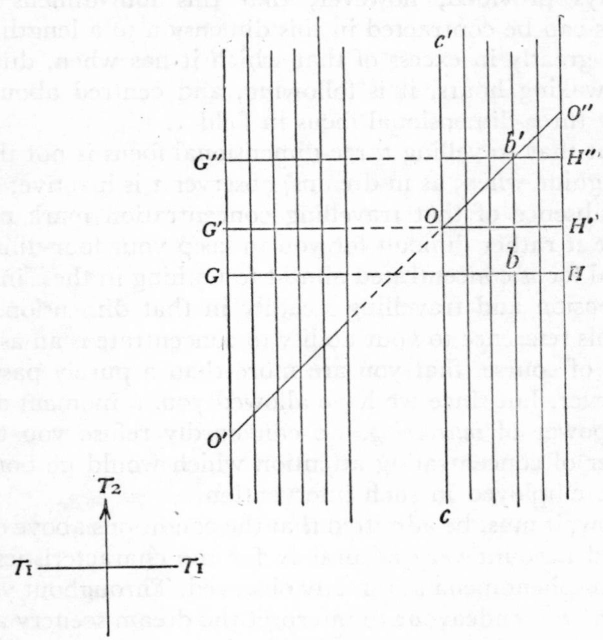

Fig. 11.

presentations which cover *periods*, and not merely instants, in Time 1. (For this dreamer, of course, Time 2 is ultimate Time.) These field 2 presentations comprise the sensory phenomena, memory phenomena, and trains of associative thinking pertaining to your ordinary waking life, but all appearing as extended—more or less, according to the degree of concentration of your focus—in Time 1. The substratum to be observed is, as always, stationary. The appearance of movements proceeding in the three dimensions of Space can be produced in the same way as it is produced in field 1 when waking—*i.e.*, by the movement of the focus of attention in the same Time 1 direction—always provided, however, that this four-dimensional focus can be contracted in this dimension to a length not very greatly in excess of that which it has when, during the waking hours, it is following, and centred about, a truly three-dimensional focus in field 1.

But that travelling three-dimensional focus is not there as a guide when, as in dreams, observer 1 is inactive; and the absence of that travelling concentration mark must make it rather difficult for you to keep your four-dimensional focus concentrated almost to nothing in the Time 1 dimension and travelling steadily in that dimension.

This reference to your ability to concentrate is an assertion, of course, that you are more than a purely passive observer; but since we have allowed you, a moment ago, the power of *intervening*, we can hardly refuse you that power of concentrating attention which would be bound to be employed in such intervention.

Now, it must be admitted that the conditions above described account very accurately for the characteristics of dream-phenomena as directly observed. Throughout your dream you endeavour to interpret the dream scenery as a succession of three-dimensional views similar to those which you experience in field 1. And always the excessive Time 1 length of your focus defeats you. Nothing stays fixed to be looked at. Everything is in a state of flux. For always your view comprises the just before and the just after of the instant of Time 1 sought for. And, because of the continual breaking down of your attempts at maintaining a concentrated focus, the dream story develops in a series of disconnected scenes. You start on a journey . . . and find yourself abruptly at the end. You are always trying to keep attention moving steadily in the direction to which you are accustomed in your waking observation —*i.e.*, forward in Time 1—but always attention relaxes, and, when you recontract it, you find, as often as not, that it is focused on the wrong place and that you are re-observing an earlier scene in the dream story. You begin to follow up what you would recognize, were you awake, as a train of associated images; but your attention relaxes slightly in the middle of the journey, so that what is actually perceived may be the first image in the train followed instantly by the last. That you seem to enter houses without passing through the walls is, of course, one of the most commonplace of happenings in a four-dimensional world.

It is very seldom, however, that you have a perfectly unbroken sleep. The brain stirs, every now and then, to a random current of nervous energy—which means that field 1 comes upon something observable. Forthwith, attention (1, 2, and the rest of them) is focused at the spot, and, as attention 1 fades again, there appears among the dream-images the four-dimensional image of which the field 1 image has just been the centre, field 1 having moved on again to a blank space. The proceeding here is precisely the same as that which occurs at the moment of falling asleep. Bodily feelings, such as pain and cold, which make themselves felt in field 1, are, moreover, confused with the true dream-images, as attention in field 1 comes into and goes out of existence. If attention to such experiences persists, you discover that you are awake.

It is a remarkable fact, however, that you never find pain or any acute bodily feeling mingling with the dream-images unless you are actually experiencing such feelings in field 1 *at that very moment of absolute Time*. And this despite the fact that your attention is travelling among brain-states, past and future, in which bodily discomfort was, or will be, distinctly present to you when awake.

The reason of this may not be far to seek. It is a well-known fact that intensity of bodily feeling depends very largely upon degree of concentration of attention. The soldier in battle often does not know that he has been wounded; you are unaware of toothache when you are running a race; attention to a bad pain will cause a smaller one to *vanish*. While, if you concentrate attention on even a very minor discomfort, this waxes until it becomes almost unbearable. Now, in the absence of the travelling three-dimensional focus of field 1 as a mark, all the other foci of concentric attention become, on our present supposition, less concentrated. Hence, in dreams —the true dreams of unbroken sleep—you are never dazzled by bright suns, deafened by loud noises, irritated by uncomfortable garments, scorched or frozen or fatigued. Dreams, although they seem real enough, lack all these unpleasant intensity-characteristics of waking life; we are barely aware of the presence of our bodies.

Pain, of course, is, according to the modern view, a sensation as distinct from other sensations as are light and sound. It has a separate neural apparatus of its own, and must not now be confused, as in the past, with that feeling of *discomfort* which accompanies the over-stimulation of sensory organs of other kinds. Pain in the eyes is something different from exceptionally brilliant light. The modern view may be expressed by saying that pain is the most disagreeable of sensations rather than that it is the sense of disagreeableness. Like all other sensations, its range of experienceable intensity must be limited. One cannot perceive colours down to an unlimited degree of dullness, or up to an unlimited degree of vividness. That one does not experience pain of less than a certain degree of intensity is obvious to any experimenter; that unconsciousness intervenes when the intensity of that sensation rises to a certain limit was the outstanding difficulty of the medieval torturer. Pain's extreme unpleasantness, and the fact that it partly distracts attention from other sensations, does not mean that this *range* of observable intensity, from the just perceptible to the absolutely unbearable, is a *long* one. Certainly it is not a range which, like that of colour, contains a great number of separately distinguishable degrees. The fact, then, that pain is not apparent at all to an observer using the relaxed field 2 focus of 'dreamland' may mean merely that the *range* of observable intensity pertaining to this unpleasant and overbearing phenomenon is considerably shorter than the range which pertains to the observable intensities of the sensation of light.

Now, throughout your dream, you *think* about that dream, just as you think about your sensory experiences in waking life. You estimate the significance of what you see in the dream; you make naïve plans to cope with the dream situations; you remember what has happened immediately before in the dream. And this is that additional thinking and remembering which we are trying to examine.

It would be going too far to say that it is, in every sense, the thinking of a little child, for it involves conceptions which pertain to adult life—such as, for example, political ideas. But we may all admit that it is thinking of an extraordinarily feeble kind as compared with that which accompanies the inspection of the successive brain-states in field 1. Yet it is, very clearly, thinking of the same general character as that of our waking speculations. It is, as we have seen, based upon the idea that the perception of a succession of three-dimensional aspects is the only possible method of observational experience; it ignores the little before and the little after of the Time 1 instant sought for, regarding this as being mere instability in what is observed; it memorizes what is past in the dream in the same would-be-three-dimensional fashion; and it causes attention, when concentrated, to travel in the *accustomed* Time 1 direction, despite the fact that Time, for the thinker in question, is at right angles to that dimension.

It is true that one does not ascertain all this from observation of the dream, but from observation of the memories of the dream, after waking. But it is not observer 1 who is inspecting those memories. *They are not in his field.* Such remembering, when awake, of what you saw in the dream and of how you thought about it during the dream is something which you accomplish without the assistance of observer 1.

Let us consider here the imaginary case of a purely automatic observer 2 whose remembering and thinking were completely analogous to those of our first-term observer. This supposed super-individual would be equipped with memory traces extending in an associational network at right angles to Time 2. His thinking would consist of the wanderings of attention over this associational plexus—wanderings to and fro in Space and backwards and forwards in Time 1. It would be thinking of a glorified, four-dimensional kind, in which Time 2 would be the only apparent Time dimension, and in which the four-dimensional way of regarding the substratum would be the natural and obvious way. This observer might be aware that all four-dimensional things were composed of an infinite number of three-dimensional sections; but he would never perceive, or try to perceive, as we do in dreams, one of those sections as unique, and the remainder as unstable, confusing additions.

Now, the records of the wanderings of the real observer 2's attention in dreams—the records which enable you to remember those dreams—*must* be traces extending in four dimensions (Time 1 and the three ordinary dimensions of Space). And, whether these traces be in the cerebral substratum or in the Time-travelling observer 2 (who is a four-dimensional entity distinct from the substratum over which he moves), or anywhere else, they are bound to constitute some sort of an associational network.

So we are confronted with the case of an observer who actually does possess the mental *structural equipment* adapted to the viewing of presentations in their four-dimensional entirety, but who endeavours, nevertheless, to regard such presentations as merely three-dimensional phenomena.

Your thinking, in the absence of observer 1, involves, therefore, something over and beyond the mere *inspection* of a four-dimensional associational structure. It involves *interpretation* of that structure.

[So it begins to look as if Professor W. McDougall were right in one main particular. For nearly all his arguments in favour of the existence of the animus amount to an insistence that what he calls '*meanings*' are interpretations by the animus of what is presented in the way of imagery by the brain. Yet it would be difficult for us to accept McDougall's view in its simple entirety. There is an opposition theory too strong and too eminently reasonable to be ignored. It is, I think, best expressed by Professor J. S. Moore, who declares that 'Meaning is *context*', and proceeds to argue that the meaning of a specific idea is simply the fringe of associated ideas which constitute that context.

The answer given by Serialism seems to be that Moore is right, but that McDougall, nevertheless, is not wholly wrong.

If meaning is given by context—by attendant associations—it must be given by the fringe of a partially relaxed attention. And this is borne out by the fact that, when our attention to an object is greatly concentrated, we notice the quality and form of that object at the expense of noticing its meaning. Now, the attention of observer 2, when surrounding and following that of a waking observer 1, is, on our theory, kept concentrated in the Time 1 dimension; and changes in concentration take place mainly in the three dimensions of Space. So that contexts, to the waking observer, are mainly relations of spatial position and spatial motion. And that is certainly true of the meanings which he attaches to what he perceives. The contexts supplied by the very slightly overlapping fringe of attention in the fourth dimension are those which exhibit the Time-travelling of observer 1, and a hint of Time 1 *pattern* in the substratum.

All of which fits in very nicely with Moore's definition.

But to our imagined automatic observer 2, thinking—in the *absence* of observer 1—in four-dimensional fashion, contexts in the fourth dimension should be interpretations as clear as are those in the three dimensions of ordinary Space. Yet it is just these fourth-dimensional contexts which are not, to the real observer, clear interpretations. And they are not clear—to him—because they are *themselves misinterpreted*—by him. Instead of being regarded as fourth-dimensional associational extensions, they are regarded as perplexing *three-dimensional instabilities*. And backward travellings of attention, from the future to the past of Time 1, are simply not noticed at all. Interpretations of that kind must be interpretations by the *observer* of the context fringes concerned.]

Here an analogy may be of service. Consider a child who, through a certain amount of experience in reading two-dimensional sheets of printed music, has acquired the habit of interpreting those sheets as arrangements of one-dimensional chords to be followed by attention in succession from left to right. When reading such a sheet he is in the position of an observer employing field 1. To extend the analogy so as to exhibit him in the position of an observer during sleep, we should have to imagine him equipped with a focus of vision which could not be concentrated enough to admit of its containing one chord only at a time. But we can get over that difficulty by supposing him, now, to be provided with a sheet in which the chords, instead of being clearly separated, are so crowded together that each partly interlocks with its immediate neighbours to right and left, the result being that no chord can be seen singly by itself. Now, none will deny that the child, presented with such a sheet, would begin by trying to read the puzzling thing in the old, accustomed way, or that the habit which compelled him to this would be, not in the sheet, but in his mind. So it is that the habit of three-dimensional interpretation which afflicts us in dreams is not a feature of the four-dimensional phenomena observed, but a characteristic in ourselves as observers. As for our inability to notice in dreams the movements of our attention backwards in Time 1, the habit of interpretation established in the ultimate thinker is amply sufficient to account for this. No child, reading a sheet of music, observes what his eyes pass over when he moves them back to the beginning of a new line. You (I hope) have read every word from the beginning of this book, and your gaze has flashed back thousands of times from the right edge of the page to the left; but never once have you read a line backwards, or even noticed what the backward aspect of a line looks like. In fact, even now that you try, you cannot perceive that aspect; and the nearest approach to a realization thereof that you can achieve is that which you obtain by viewing a word written backward, but still from left to right—looking-glass fashion. And the habit which blinds you to that aspect is not in the printed page, but in yourself.

So we are driven to the interesting conception of a higher-order thinker who is *learning to interpret* what is presented to his notice, the educative process involved being his following, during the waking hours, with unremitting, three-dimensional attention, the facile, automatic action of that marvellous piece of associative machinery, the brain.

This, admittedly, is a complete reversal of the old-time animist's conception of the 'higher' observer as an individual of superlative intelligence producing the best effect he can with the aid of a clumsy material equipment. But it seems to me there is no getting away from the plain evidence afforded by the character of our dream thinking. Whatever capacities for eventually superior intelligence may be latent in the higher-order observer, they are capacities which await development. At the outset brain is the teacher and mind the pupil. Mind begins its struggle towards structure and individuality by moulding itself upon brain.

Evolution has worked for possibly eight hundred million years towards the development of brain. To-day, as Professor McKendrick points out, nearly all the functions of our bodies are operating towards the end of the adequate nutrition of the *grey matter*. And it now appears that, apart from its self-sustaining and self-developing activities, the brain serves as a machine for teaching the embryonic soul to think.[1]

[1] Reasoning is a retrospective business—the judging of a present situation in the light of past experience. Intuition is more akin to the simple inspection of a field 2 pattern. The former process, employed at the expense of the latter, is the concomitant of any life of adventure.

We are now in a position to consider what is the *origin* of the habit which keeps the higher-order observer's attention focused in field 1.

In field 1 he has to deal with merely a simple succession of three-dimensional phenomena in a three-dimensional field. But in field 2 he is confronted with a view of four-dimensional phenomena in a four-dimensional field. And, in addition, he has these four-dimensional phenomena duplicated. For example, he may find at *a* (Fig. 10, page 163) a memory revival of a preceding event in Time 1. And he has also, somewhere between *G* and *a*, the original event which originated the memory traces subsequently revived. In field 3 the substratum (see Fig. 9, page 156) is crowded with five-dimensional phenomena (containing, however, none that are not already represented in simpler four-dimensional form in field 2); and these phenomena, owing to the less concentrated area of the focus of attention, are of less intensity than are those in either field 1 or field 2. And the intelligibility of the presentations gets worse, and their vividness gets less, as we proceed up the series.

It is in field 1, then, that, for the infant, phenomena first become *distinguishable* at all. And his attention stays where there is something to be attended to.

Next, we know that, even within the limits of field 1, an adult's attention may be attracted from without as well as directed from within. We know, also, that the directing of attention away from a point of attraction is a process which has to be learned, painfully, at the schoolroom desk. The young child's attention must be, therefore, largely at the mercy of attraction. And we know that the greatest attractors of attention are the cruder bodily pleasures and bodily *pains*. These exist only in field 1. Thus pain performs a service other than purely physiological.

Finally, the child learns quickly enough that in field 1 he can *intervene* to obtain those pleasures and avoid those pains. And that, very rapidly, becomes the dominating aim of the man.

* * *

Reviewing the foregoing parts of this chapter, we see that Mind—the Mind which can appreciate only the most elementary aspects in the complex structural equipment at its disposal—must always exhibit itself as something external to any *structural* conception thereof that we can attempt to form.

* * *

In Part I of this book we carefully refrained from tackling the question as to whether the internal *directing* of attention was to be attributed to the higher-order observer, or to be regarded as originating in the purely automatic internal condition of the brain. We contented ourselves with noting that, if we regarded the higher-order observer as the responsible agent, we should be according him the status of an animus, with power of intervention, since the concentrating of attention is known to have a marked effect in the formation of memory traces.

It would be best, however, in order to avoid any possible trap for an incautious thinker, to show that such directing of attention—such intervention—must be attributed to the higher-order observer.

The question is, really, whether, in any higher field, attention may be bound to coincide with some feature in the substratum analogous to the 'maximum flow of cerebral energy' in field 1.

We saw, earlier, that the analysis had brought to light no law which *compelled* attention to direct itself upon any particular phenomenon in any particular field. It was pointed out that attention, which is referable to the higher-order observer, was sharply distinguished by the analysis from that which was presented to attention; that is, from the contents of the substratum. Now, 'maximum flow of cerebral energy', or anything analogous thereto in any higher field, is a substratum feature, and, as such, categorically distinct from 'focus of attention'. Theoretically the two things may be separated. And that this theoretical distinction is a practical, real distinction, and not merely a bit of metaphysical hair-splitting, is shown by the 'waking experiment'. For there the one thing is present and the other is absent.

There is one great difference between the conditions in this waking experiment and those which obtain in dreams. In the former case the cessation of field 1 attention, which sets free field 2 attention, *is not accompanied by the cessation of body-maintained cerebral activity*. The eyes may be open, transmitting to the cerebrum light-stimulations differing in intensity at different parts of the field of vision. Noises of various degrees of loudness are assailing the ears. Cerebral action is flooding associational tracts, presenting those hosts of associated images to which attention (this, as we saw, is the very essence of the waking experiment) must be determinedly refused.

This shows that the theoretical distinction between the focus of attention of the higher-order observer and any line in the substratum which it may habitually follow is a real one, and so we are bound to regard it as always *possible* for such focus to be separated from any such line. And, where the two things do coincide, the higher-order observer must be regarded as an accessory, passive or active, to that coincidence.

All of which, of course, is to admit that the higher-order observer is an individual potentially capable of exercising what is called, rather vaguely, '*freewill*',[1] though how far he may be said to have developed that capability is quite another matter.

That he can, and does, direct attention in field 1 is now plain enough. But his control in field 2 seems to be as limited as is his comprehension of that area. We may note, however, that, throughout his dreams, his rudimentary intelligence is extremely active in attaching interpretations to that which he observes. (Indeed, as I remarked earlier, he is a master-hand at attaching wrong ones.) And it is a matter of common knowledge that he employs his function of interpretation in weaving a dream *story*—a drama of personal adventures—out of the various presentations upon which his attention becomes focused. If he can direct his attention at all in this field, he can modify the trend of that story; can, in fact, build the drama to please himself. He has an immense wealth of material. He is, as we have seen, potentially capable of exercising that control, and, judging from my own experience, I am disposed to think that he does do so to a small extent, and that his effectiveness in that respect increases with practice. Adults, I fancy, are not so much at the mercy of their dreams as are children; they can (certainly I can), occasionally, alter a situation which fails to please.

These, however, are matters for the psychoanalyst. But perhaps when we have learned to interpret fourth-dimensional contexts as 'present' wholes—to think four-dimensionally—and to master the movements of our attention, we may find field 2 of greater interest than field 1. But that development in comprehension and control is not likely to occur so long as we continue to spend nineteen hours out of the twenty-four in practising attention to the experiences of observer 1.

[1] Nobody means by 'free' will a thing actuated by no motive whatsoever. But the motives of observer 2 may be, in some circumstances, flatly opposed to those of observer 1. There is a very obvious 'regress' here.

We must live before we can attain to either intelligence or control at all. We must sleep if we are not to find ourselves, at death, helplessly strange to the new conditions. And we must die before we can hope to advance to a broader understanding.

CHAPTER XXIV

Consider, now, the situation represented in Fig. 12. When (in absolute Time) field 2 is at *GH*, the substratum between *a* and *H* comprises an ordered arrangement of three-dimensional cerebral states, all in the future part of Time 1. That thinker who is the observer of dreams —which involve the observation of field 2 as present— observes, let us say, at that moment, one of these future states *b′*. After waking, when field 2 is at *G′H′*, and field 1 is at *O*, this thinker intervenes at *O*. That intervention we will suppose to be due to his memory of the dream; just as every word which I write in this book is intervention due, originally, to my memories of similar dreams. (The diagram, however, will serve equally well to illustrate the results of an act of intervention originating in any other activity of this observer's partially trained mind.) Now, we have to note that such an act of intervention may result in the complete alteration of part of observer 1's future career. Taking the train to Dover instead of the express to Southampton may lead to his being decapitated by Russian politicians instead of being clubbed by a New York policeman. So he may never encounter the cerebral event represented by *bb″*—the event perceived in the dream— and may, instead, when field 2 is at *G″H″*, encounter a totally different event, *c*.

In the sort of life led by the average civilized man, intervention has seldom any very great effect in altering future experience. We live too much in ruts for that. A man may, on Monday, take a ticket for a Saturday matinée, and he may, during the next few days, perform countless little acts of intervention; but these will not necessarily prevent his occupying his seat on the Saturday, or prevent his seeing on the stage a scene of which he may have dreamed on Monday night. The intervention at *O* may, thus, alter some of the events between *O* and *H'* while leaving others unchanged. In fact, if we represent the alterations by breaks in the vertical lines, just above *OH'*, the result would be the sort of thing shown in the figure.

It is to be noticed, however, that these breaks in the verticals are to be regarded, not as fixed substratum features which exist before (in absolute Time) observer 1 reaches *O*, but as *changes* in that substratum which occur at the instant when (in absolute Time) this observer reaches that point. This means that the breaks are being represented as due to intervention, and *consequent* upon the higher-order thinker's interpretation of the event which he has, in his dream, perceived at *b'*. (We saw in the last chapter that such interpretation cannot be represented as any sort of context or trace in the substratum.) To regard the breaks as pre-existing (in absolute Time) fixities in the Time-map travelled over would mean that the higher-order thinker would encounter the new event whether he had the dream of the old one or not: the breaks would not be occurring as the result of the *dream*.

We saw in the last chapter that all movements of attention require passive consent or active intervention on the part of the higher-order observer. Where such movements involve a departure of attention from that line in the substratum which represents the flow of maximum cerebral energy, we have active intervention accompanied by substratum changes similar to those shown in Fig. 12. But, considering the degree of intelligence which the intervener exhibits when the brain is dormant and not employable as an aid to his reasoning, we cannot conceive that his interference with cerebral thought processes amounts to very much more than an insistence that the machine in question shall operate towards a certain end of his own. The intervener, in fact, is analogous, not to a skilled musician composing with the aid of a piano, but to the amateur user of a pianola, whose interference with the

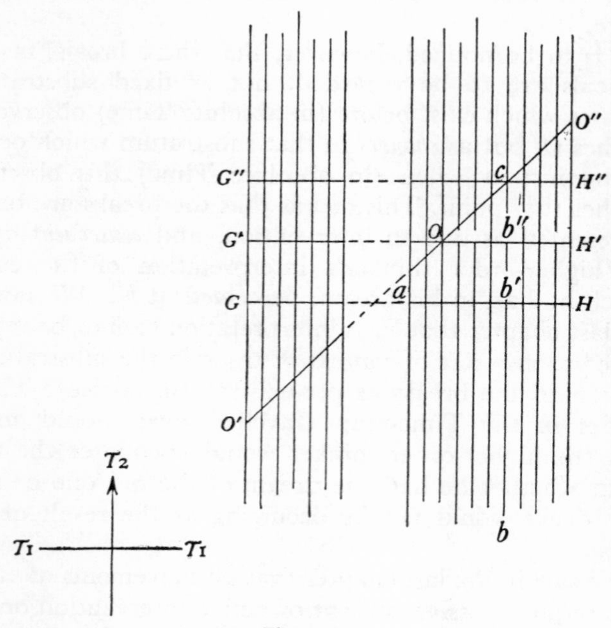

Fig. 12.

complex performances of that instrument is limited to the changing of one perforated roll for another.

That the change in the substratum takes place all along *OH'* instantaneously (in absolute Time) is obvious enough when we regard the effects of the intervention from the standpoint of our more customary, three-dimensional philosophy. None can deny that, when he takes a step to prevent an otherwise probable event from occurring, the probability of that event (however distant) being encountered is altered at the precise instant when he takes that step. Translating that into the language of four-dimensional philosophy, it means that the probability of observer 1 encountering the event *bb″* when (in absolute Time) he arrives at *c* is changed at the 'precise instant when' he intervenes. That 'precise instant' is an instant in the Time which times his travel along *O′O″*, which Time is Time 3—the absolute Time for the diagram. The breaks occur, therefore, when (in absolute Time) observer 1 reaches *O*, which is when (in absolute Time) field 2 reaches *G′H′*. The altered course between *O* and *H′* will be, in all its parts, a mechanical sequence just as perfect as before.

[It is clear that the alteration of the substratum along *OH′* must affect also that extension of that line as a plane (perpendicular to the paper) which represents the line's endurance in Time 3. The 'future' part of that plane must change with the change in the line. And so on through all the futures ahead of *O* in all dimensions of Time. Consequently, nowhere in our serial Time-maps can we pick out a path *ahead* of *O* which is absolutely assured in all its parts.]

* * *

It is essential that we consider the series so far as to include the second term, otherwise the serial relation will not be fully disclosed. But there is no practical object to be achieved by considering the remoter terms. You will find that no new kinds of relation between observer and observed become apparent in the third term. Carrying the analysis further means merely pushing back the higher-order observer and thinker, with all his peculiar functions, and the insertion of additional reagents, all reacting to the contents of the substratum, and all unconscious save where this observer employs them to gain an acuter view.

It is sufficient, then, for you to picture the world as containing observer 2; that is, as the field 3 which is Fig. 8. *This gives you the complete serial relation.*

CHAPTER XXV

Since observer 2 sees what lies 'ahead' (in Time 1) of observer 1, there must be something there for him to see. And whatever is thus positively there must be 'pre-determined' from the point of view of observer 1 employing Time 1. So the question arises: What is this 'something', and how is observer 1 able to alter it by intervention at his Time 1 'now'?

In previous editions and reprints of this book that question was answered in a chapter devoted to modern physics and following the lines of Sir Arthur Eddington's identification of 'probability' with the physical, four-dimensional quantity known as 'action' (i.e. energy multiplied by time). Recently, however, Serialism has invaded the realm of physics in more definite fashion, with the result that the foregoing explanation has proved capable of considerable simplification. But for a full account of these newer developments I must refer the present reader to my book, *The Serial Universe*, published by Messrs. Faber and Faber. It will suffice, I think, if I explain here that the general theory, supplemented by considerations which have been the subjects of later lectures, stands now as follows.

Any world which is described from observation must be, as thus described, *relative* to the describing observer. It must, therefore, fall short of accordance with reality in so far as it cannot be thought of, by anyone who accepts the said observer's description, as capable of containing that observer. Consequently, you, the ultimate, observing you, are always outside any world of which you can make a coherent mental picture. If you postulate the existence of other observers making different descriptions, then it turns out that *you and these other observers must together form a composite observer who is not includible in the world as thus conjointly described*. You, as part of that composite observer, retain your individuality.

The world of psychological phenomena which you describe as 'sense-data' positioned along Time 1 would be described by the postulated other observers as physiological happenings in your brain. But you and they are, severally and conjointly, not includible in any world which you and they, severally or conjointly, can describe—such as a world of brain-organisms.

The picture you draw shows the real world in its relation to yourself—shows, that is to say, how that world is capable of affecting you. If drawn as the composite effort of many observers, it shows how the physical world is capable of affecting Mind in general. The most important fact which emerges is that you prove to be the immortal part of an immortal composite observer—an aspect of the matter which we shall discuss in the next chapter.

The physical world which you (or you with others) describe exhibits itself as deterministic in Time 1. But it proves to have a contact point with your observer 1 at the travelling Time 1 'now'. There you can interfere, and every scientific experiment is such an interference. This is rendered possible by the existence of that very curious quantity known to modern physics as the '*Quantum of Action*'. Consequent upon any such interference, the Time 1 stretch ahead of observer 1 is altered; but, thus altered, it forms again a deterministic sequence starting from the point of interference. When your travelling observer 1 arrives at that place in the four-dimensional substratum where what others describe as your brain separates into its component parts, your chance of intervention ceases, in absolute Time, to exist.

CHAPTER XXVI

It is to be feared that the observer's power of interference does not suffice to make him wholly master of his fate. For there are other observers, employing similar capabilities. While our friend is in bed, dreaming of the happy probabilities of his future, some enemy, afflicted by this mania for intervention, may proceed to fire the house and reduce those probabilities to might-have-beens. (They would remain always, of course, entities in the substratum past of Time 2, but entities never encountered by field 1.) And, if the observer may owe his time 1 end to the intervention of other observers, it is fairly certain that he owes his beginnings to nothing else. Before his birth he can be nothing but a probability in the future of the race.

This brings us to the question of how the fields of different observers are related.

Our knowledge that such observers can intervene helps us to see that their respective field 1's must, in their motions along Time 1, keep within certain limits. For, if the field of an observer *A* lagged so far behind that of an observer *B* as to permit of *A*'s intervention affecting *B*'s substratum at a point *behind* *B*, then *B* would find his experiences in *his* field 1 miraculously altered. In fact, he might find himself miraculously dead, having been slain by *A*, unknown to himself, some little way back. And that sort of thing does not happen in our experience.

Suppose, now, that we were to draw a plane diagram of the 'family tree' of the entire human race, employing one dimension of the paper as Space and the other as Time 1. The result would be a network with numerous points of intersection representing marriages, and numerous branchings-off representing births. And you would find that you could trace in that network an unbroken connection between any two points that you chose to select; human families are all related in that fashion.

If we were to assume that this diagram exhibited only the *cerebra* of the individuals concerned, it would be the first, stage 1, temporal extension in a Time analysis in which we were dealing with all human observers together, instead of with one alone.

Here we may glance at a rather interesting question. Is this *network*, with wide Space-gaps between its lines, the nearest approach to a universal field 2; or is there a field 2 which fills all Space, including those gaps?

Consider again the network of this universal, 'family tree', cerebral substratum, a portion of which we may suppose to be exhibited, in perspective, by the connected lines *AB*, *BC*, and *BD* in Fig. 13.

These three lines will endure upwards in Time 2 in the forms of the planes *AA′B′B*, *BB′C′C*, and *BB′D′D*. If we ignore Relativity considerations we may say that these planes will be intersected by the respective reagents *AE*, *EC′*, and *ED′*, and also by the respective field 2's (shown at the top of the figure, for simplicity) *A′B′*, *B′C′*, and *B′D′*, constituting the portion of the field 2 network *A′B′C′D′*. Now, we know that the lines of the individual observer 2's must conform to the shape of the substratum network. If, for example, when the horizontal Time 2 plane is at *E*, intervention at that point alters the trend of the substratum, so that *B′C′* and *B′D′* depart from one another at a narrower angle than do *BC* and *BD*, then the lines of the individual observer 2's must close up to agree. But an individual observer 2, be it remembered, is *not* the substratum contents of his field

The analysis has shown that he is an independent entity, who observes those substratum contents. Why, then, is he tied to them through all their spatial windings and through all their interventional changes in spatial position?

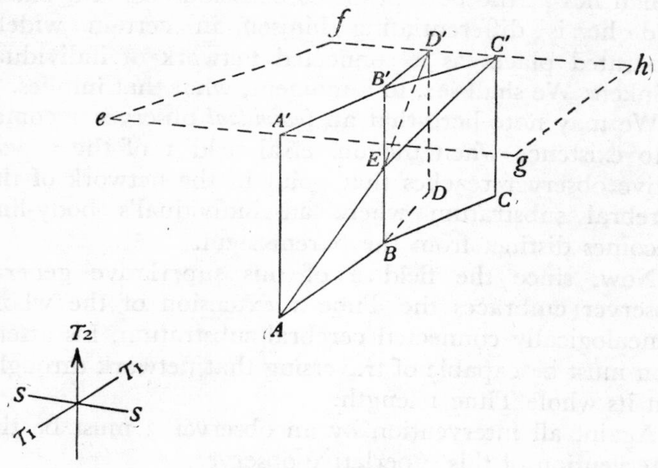

Fig. 13.

This can be accounted for only by regarding the individual observer 2's as the intersections of the substratum with a universal, Space-filling observer 2 possessing a universal field 2 similar to the plane *efhg* in Fig. 13. And the places of intersection between this universal observer 2 and the various reagents in the figure must constitute the individual field 1's.[1]

[1] In the plane diagrams of the hyperbolic world of Relativity the individual observer 1's do not lie on any common-to-all straight line except when their 'tracks' are parallel—which last is practically the case so far as the denizens of this planet are concerned. But in all cases these observer 1's are distributed within the transverse 'hour-glass' area between the light lines. This area travels through the map, the question of the direction of its travel introducing the second-term Relativity.

Now, we have seen that the higher-order thinker in the series pertaining to each individual observer is learning to think in terms of mechanical brain-thinking. So, if we halt at this stage, the universal observer must be, throughout his Space-filling area, the unknown element which lies at the bottom of self-consciousness and mind, and he is differentiating himself in certain widely separated places as a connected network of individual thinkers. We shall see, in a moment, what that implies.

We may note here that an *individual* observer 1 comes nto existence when the universal field 1 of the superlative observer reaches that point in the network of the cerebral substratum where an individual's body-line becomes distinct from the parent stem.

Now, since the field 2 of this superlative general observer embraces the Time 1 extension of the whole genealogically connected cerebral substratum, his attention must be capable of traversing that network throughout its whole Time 1 length.

Again, all intervention by an observer 2 must be the intervention of this superlative observer.

We may sum up, therefore, by saying that this superlative general observer is, at this stage, the fount of all that self-consciousness, intention, and intervention which underlies mere mechanical thinking; and that he, in his intersections with the cerebral substrata, is incarnate in all mundane conscious life-forms, in every dimension of Time; and that he must—owing to the unity of the network thus formed in himself and the ability of his attention to range over that network's full extent—contain in himself a distinct *personification* of all genealogically connected conscious life—a Synthetic Observer. And we may add that this 'personification' must be capable of thinking on a scale rendered ampler than ours by the immense Time 1 range and Space range of his field 2, and by the immense age of his experience as a thinker in that field.

We have wandered from our main task into what appears to be a region for exploration by the theologian. Let us leave it to him (he will find an extraordinary number of *dicta* which fit the case), and get back to our proper business.

This book is not intended to be anything more than a general introduction to Serialism as a theory of the universe. Every such theory must have its psychological, its physical, its theological, and its teleological aspects. At each of these we have glanced briefly, yet long enough to show us how large and how promising is the field for investigation opened up by the new method of analysis. But exploration proper in these several regions has been regarded throughout as the province of specialists more directly concerned.

The man-in-the-street, however, will expect something in the nature of a summarized statement as to how he is to regard Serialism as affecting *himself*. Such statements are not always advisable—for reasons which will be clear enough to the judicially minded. But, in the present case, all the points which do directly affect the man-in-the-street have had to be touched upon in the course of the book, since it so happened that none of these points could be omitted from consideration without breaking off the argument at a critical place and leaving the theory, so to say, in the air. There can be no harm in summarizing in one place what has already been said in odd paragraphs throughout preceding pages.

Putting it roughly, then, I should say:

1. Serialism discloses the existence of a reasonable kind of 'soul'—an individual soul which has a definite beginning in absolute Time—a soul whose *immortality, being in other dimensions of Time, does not clash with the obvious*

ending of the individual in the physiologist's Time dimension, and a soul whose existence does not nullify the physiologist's discovery that brain activity provides the formal foundation of all mundane experience and of all associative thinking.

2. It shows that the nature of this soul and of its mental development provides us with a satisfactory answer to the 'why' of evolution, of birth, of pain, of sleep, and of death.

3. It discloses the existence of a superlative general observer, the fount of all that self-consciousness, intention, and intervention which underlies mere mechanical thinking, who contains within himself a less generalized observer who is the personification of all genealogically related life and who is capable of human-like thinking and prevision of a kind quite beyond our individual capabilities. In the superlative observer we individual observers, and that tree of which we are the branches, live and have our being. But there is no coming 'absorption' for us; we are already absorbed, and the tendency is towards differentiation.

4. Its proof of the unity of all flesh in the Superbody and of all minds in the Master-mind supplies the logical foundation needed by every theory of ethics.

5. It accounts for dreams; it accounts for prophecy; it accounts for self-consciousness and 'freewill'; while, in its disclosure of the relations between the general and the individual fields of presentation, it provides the first essential to any explanation of what is called, loosely, 'telepathic communication'.

6. It does not contradict either modern physics or modern physiology.

A theory which can achieve all this is not lightly to be set aside.

PART VI

REPLIES TO CRITICS

An Experiment with Time was published in 1927. Its reception by men of science has been generous to the point of indulgence, and the theory has been the subject of unexpectedly continuous attention by the public press. But, in the world of metaphysics, there are still some who regard it with grave suspicion. This last is hardly a matter for wonderment, when we recall that the archaic rules of non-mathematical philosophy allowed that victory might be granted to any disputant who could prove that he had driven his opponent into the gateway of an 'infinite regress'. For here is a book which asserts flatly that philosophy has been at fault, and that an infinite regress is, after all, the proper and valid description of mind's relation to its objective universe.

The problem for these particular philosophers has been that they wish to accept the evidence for dream precognition, while continuing to teach that every infinite regress must be 'vicious'. Obviously, they are attempting the impossible. The effort, however, has been made, so I have tried to describe it below. But the reader must not blame me if I have failed to make it sound plausible.

First, a four-dimensional observer, with a 'present' four-dimensional outlook and with time as a fifth dimension, is accepted. (That is inevitable, since dream precognition is granted.) But the admission of this observer is made on purely experimental grounds, and no regress, so far, has been acknowledged. Next, we have to account for the also empirical fact that we perceive as a rule, when we are awake, a simple succession of apparently three-dimensional events. It is to be noted, of course, that, if the accepted four-dimensional observer were to travel in the fourth dimension with the velocity of light, this would cause his fourth-dimensional length to shrink to nothing, and, so, would reduce him to a travelling three-dimensional observer marking a Time 1 'now'. (That 'now', as shown in *The Serial Universe*, travels with the velocity of light.) But then, in order to achieve the four-dimensional outlook evinced in dreaming, he would need to slow down very considerably his rate of fourth-dimensional travelling. And the objections to a single observer with a variable velocity of this kind are insuperable. On waking, he would have to spurt forward with a speed immensely greater than that of light, in order to catch up with other observers who had stayed awake. But he would attain to the speed of light (would become a normal, three-dimensional observer) *before* overtaking those others, and his interferences at that instant would produce in many circumstances for those others effects which they would regard as staggeringly miraculous (*i.e.* he could alter their pasts and, so, change their entire present conditions). The concept of a single observer travelling along the fourth dimension at the velocity of the 'now', and expanding his focus of attention when sleeping, so as to cover part of the fourth-dimensional stretch ahead of him, is prohibited by the fact that his rate of travel would keep him reduced to three dimensions and prevent any such forward expansion of attention. (And it is to be noted that, from the psychological standpoint, attention, with its *focus* broadened to cover, say, three days, would perceive nothing but a blur.) So, to account for both dream precognition and normal waking experience, we have to accept a four-dimensional observer equipped with a three-dimensional sub-observer

—a 'self' employed as an instrument. But (thus runs the argument) we can claim that both observers, the four-dimensional and the three-dimensional, have been discovered *empirically*, and that we have not had recourse to the concept of a regress of times or observers.

I can see no sense in this argument. The two observers in question, with their two times, behave precisely as do the first two observers of the abhorred series, and, consequently, may be regarded as empirical evidence of the validity of that regress. In fairness, it must be admitted that authors of the attempted compromise recognize usually the advisability of producing some argument to show that this awkward coincidence between theory and practice cannot be, reasonably, more than a coincidence; and, indeed, the necessity of doing this becomes imperative in view of the fact that Chapter XXI of this book claims to prove, logically and without reference to dreams, the existence of the serial observer. That alleged proof is a challenge to the philosopher which awaits an answer. None has attempted a refutation. Leaving aside those who ignore the argument, on the plea that they cannot understand diagrams, we are left with the assertion of the single writer who has said that, if the first step in my analysis is correct, the remainder must follow automatically, but that he can perceive no grounds for that initial departure from orthodoxy. Now, the justification demanded was given in detail in Chapter XXI, and the critic does not attempt to lay his finger on the flaw. It is as if I had cried, 'Checkmate!', and my opponent, instead of replying, 'Not at all, I can move to this square', had leaned back in his chair and remarked merely, 'I do not see that I am mated'. The correct procedure for me now is to say, 'Very well, get out of it if you can', which, of course, is merely to repeat, 'Checkmate!'; but, in view of the advisability of getting the question settled, I am quite willing to waive my rights and to show precisely how every avenue of escape is blocked.

Hinton and his forerunners, referred to in Part IV of this book, made no attempt to show in what way the elaborate conception of a three-dimensional observer traversing a four-dimensional world is superior to the orthodox method of picturing temporal phenomena. Orthodoxy represents all events, *including those which implicate the supposed observer of external successive happenings*, by points arranged in a spatial order chosen to indicate

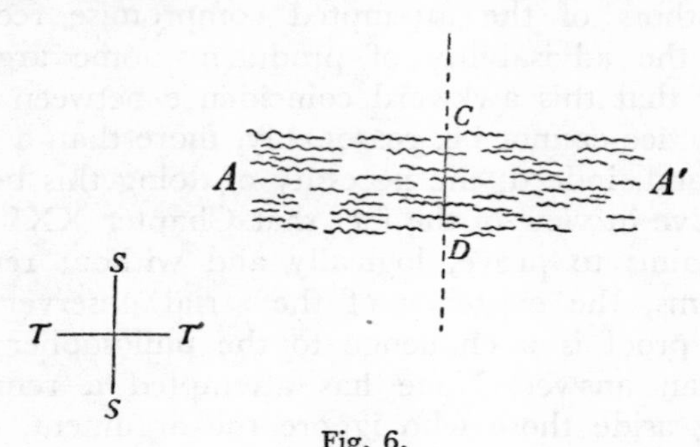

Fig. 6.

unidimensional time, and by an arrow which shows the direction or 'sense' of from earlier to later, thus distinguishing the pictured 'time dimension' from the pictured other and purely spatial dimensions in which a to-and-fro motion is possible. But this orthodox description is linked to a grave error in its treatment of the observer of succession. That was what Hinton was required to show, and that was where he failed to rise to the occasion. It was to rectify his omission that the first nine pages of Chapter XXI of this book were written.

I will ask the reader to look now at Fig. 6 (repeated above), and to consider the band *AA' with the cross line CD omitted for the moment*. My opponent would declare that this represents quite adequately the successive states of the brain of an observer whom we may call Smith. He would agree, moreover, that these cerebral states are accompanied by certain sense data. But he would deny the existence of any Smith who is not indicated already in the diagram. Smith—the only Smith—he would say, is the brain we have pictured: there is no observer of sense data: the sense data are the observations made by Smith. The reader must not expect me to make that sound reasonable—it is simply the fallacy which I have to expose.

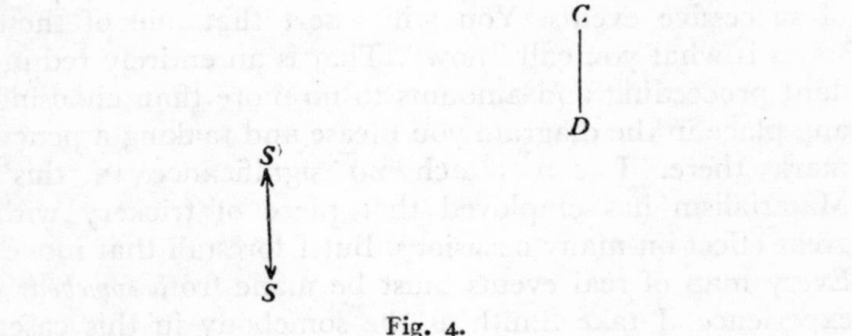

Fig. 4.

I begin by leaving the point in dispute entirely open, and I introduce, as *CD* in Fig. 4 (see above) the 'field' of the presented sense data. This field is not a thing: it is a mere mark indicating the spatial extension of the moving neural correlates concerned, and indicating also that such motion has three degrees of freedom (*i.e.* the field is three-dimensional). I have kept that field well to the front during the earlier part of the book. In particular, I would refer the reader to pages 24, 25 and 26. There he is shown how to avoid confusion between the field and attention wandering within that field. For additional security, I have re-emphasized that lesson on pages 136, 137 and 138 immediately preceding the argument with which we are concerned now, it being still left open whether the observer to whom the field pertains is or is not a creature composed of the neural correlates contained in the field.

The next step is to map out or 'extrapolate' in Fig. 5, as our necessary first attempt at a picture of time, the endurance of the *contents* of the field. And this is where I make my attack. My opponent has omitted to consider by what process the nature of the extrapolation AA' is supposed to be discovered. He would like to start with an AA' of unknown origin, *plus* (or, perhaps, even without) an arrow, and then go on to say, 'This represents a series of successive events. You will assert that one of these states is what you call "now". That is an entirely redundant proceeding and amounts to no more than choosing any place in the diagram you please and making a pencil mark there. I can attach no significance to this'. Materialism has employed that piece of trickery with great effect on many occasions. But I forestall that move. Every map of real events must be made from *somebody's* experience. I take Smith as the somebody in this case; I let him watch changes taking place in the three-dimensional field; and I let him say *when* he is going to start mapping out those changes. The fact that he *can* say when is of supreme importance. From the information available at the chosen moment, Smith prepares the map AA'. If he is a physiologist, he can himself translate any sense data that may be concerned into terms of neural correlates; otherwise, he can call in a physiologist to his assistance. He marks within the field CD the positions which the neural correlates occupy at the chosen moment when the extrapolation is to be made; he places to the left in proper time order the states of the neural correlates of all such experiences and activities as he can remember; and on the right he fits in such states as he can anticipate by logical processes. There is a little uncertainty about these prophecies of Smith's which is of great value in establishing in Physics the importance of the 'now', but that is dealt with in *The Serial Universe* and need not concern us here. The point is that, in this time map constructed by Smith, the section marked out by *CD* contains what is present to *this* Smith (the Smith who says 'when'), and sections to the left and right thereof contain states which are *not* present but are only remembered or anticipated. So, in this map, Smith, the observer of succession, can be inserted at one place only, the place marked by *CD*. And this Smith, who is not represented anywhere in the diagram, observes, remembers and anticipates sense data. It is clear, moreover, that this Smith, outside the picture, not only draws the picture with reference to a personal 'now', but must show all the states in *AA'* as streaming past that 'now', so that the state at *CD* will be replaced for him, later on in that absolute time which we have not yet succeeded in defining, by states which he has depicted to the right of that mark.

Now, the favourite supposition of materialism has been that an examination of Smith's relation to an assumed external world existing independently of Smith would show that his picture of that relation is (in some unexplained fashion) 'illusory'. A materialist would affirm, for no clear reason advanced, that an independently-made time map would show that the Smith who constructs his own map is nothing but a set of brain states between *CD* and *A'*. But I anticipated that move, and blocked that way of escape. For I called in the assistance of this independent observer at the outset of the analysis (*vide* the third paragraph of page 139). You, the reader, were asked to stand by and check the accuracy of Smith's work. You agreed with Smith as to the instant when the extrapolation was to be commenced. Now, it is clear that you must needs make a map of your own which is similar to Smith's in the respect that whatever is present to *you* is at *CD* only. In your map, your 'now', like Smith's, must be regarded as moving along Time 1. But you agreed from the beginning that Smith has a single three-dimensional field of presentation with changing contents, and, in your Fig. 4, you must place that field of Smith's at your *CD*. (Remember that you are standing beside him.) That single, three-dimensional, present field of his must, in *your* map, accompany your travelling 'now'; otherwise there will be no Smith who observes a single field with changing contents. The Smith who makes his map is, therefore, at your travelling 'now' as well as at his own.

All observers must come to the same conclusion about other observers as you have come to about Smith. Hence all observers must agree concerning the presence of their own and other persons' three-dimensional selves at 'now's' which travel along Time 1 *and which vary in their alignment only to the extent that this is permitted by or dictated by the rules of relativity.*

The upholder of the orthodox theory is, thus, completely encircled. He is checkmated because the ground to which, hitherto, he has been able to retreat without incurring from his dazed opponent the penalty of a false move has, here, been denied to him in advance by the procedure of the attack.

The regress thereafter follows, as has been agreed to by my adversaries, automatically.

I am afraid, then, that classical philosophy will have to reconcile itself to the deplorable fact that Smith, his life and the world of which he forms a part are all incurably 'vicious'.

And I may add, without wishing to indulge in personalities, that, as regards the present reader, the verdict of this book is plain. Time, as it affects him, is serial; and, in his relation to that (to him) all-important time, he is a serial observer.

* * *

The following remarks have been added at a slightly later date (January 1938).

The portion of this book which precedes Part V is the story of a detective who is collecting evidence and describing, with brief comments, the theories of other persons—theories which he hesitates to accept. Many criticisms which have been published consist in attributing to myself the theories of these other people. Broad, for example, says that I accept Hinton and carry on from there. Miss Cleugh cites an extract from Chapter XVII wherein I say, 'The employment of these references to a sort of Time behind Time is the legitimate consequence of having started with the hypothesis of a *movement* through Time's length.' But in that Chapter I am pointing out merely that most people *begin* by spatializing time and that to do this initiates an obvious regress.

In Part V the detective takes off his coat and starts to work out his own theory.

I am investigating what we mean by 'happening'. Miss Cleugh says an event is that which happens, and that to assert that the happening of an event is itself an event is a fallacy causing a regress. Weyl says that events do not 'happen'—we come across them. Now, Miss Cleugh, clearly, is wrong. An event is the happening of something, e.g., a material configuration, which is not in itself an event. It happens, and that makes it an event. To go further and say that the event happens is to introduce, herself, the regress she is trying to avoid. Weyl, also, like Hinton and the-man-in-the-street, begs the question. All these are adopting, as a supposition without previous justification, the viewpoint of observer 2. I decide that we must start earlier; and I proceed to investigate what are the attributes we should be justified in granting to time in an *objective* world, the condition being that this time should play the largest part possible in accounting for our experience of change in that which is observed.

The intuition that time has elapsed between two particular experiences is essential to any awareness of time; hence, any time concept must be the concept of a one-dimensional *continuum*. Now, if we were to begin by saying that the objective world has space intervals only, and that some of these are misinterpreted as time intervals, we should be committing ourselves to a regress (and a wrong one) before any regress is proved. We shall try, therefore, granting *time* intervals to the objective world. We shall attribute, moreover, causation to that world, in the sense that its events, separated by its time intervals, constitute a 'oneway' causal system. Do we need an external 'now'? No, for the events, the happenings of material configurations, represent a whole series of 'nows'. Let us, then, commence our picture. But how? Here epistemology asserts its rights, and we discover that we have to extrapolate from somebody's 'now'.

Smith, who extrapolates, is trying to regard the continuum in question, not as space, but as time. That it contains a series of events is sufficient for that purpose; but, to make assurance doubly sure, he declines to presuppose that the external events which are past and future to him are real and coexisting. His subsequent discovery that (1) the external events which he has labelled 'future' and 'unreal' become 'present' and 'real', and (2) that this is due to *nothing which he can embody in his descriptions of these events*, compels him to recognize (a) that there is a 'now' travelling over the continuum in question, and (*b*) that this 'now' is not an adjective but a *thing* which will have to be brought into the picture. He discovers, in short, his 'self' (Smith 1) travelling over a series of external events which have no reality distinction inherent in that series and attributing reality to these events in succession merely because it observes them in succession. That 'self's' successive coincidences with these objective events constitute a series of double events—observational events—and *this* series requires a time 2 for its representation. It becomes apparent then that what is pure time from the point of view of Smith 1 is, from the point of view of Smith 2, something which has a property of space, in that it can be moved over, and a property of time, in that it is a one-way causal system.

It has been noted that the book has not referred to the psychologist's 'specious present' under that name. It has, however, introduced this as the four-dimensional focus of attention of observer 2. Now, all students of the specious present agree that this has a mysterious centre—a culminating peak of vividness. That centre is provided by observer 1.

Many people, I hear, suppose that there is some clash between serialism and the 'wish-fulfilment' theory of dreams. There is none. 'Wish-fulfilment' theories are concerned with explaining why the dreamer builds a particular dream edifice: I am interested in the quite different question of whence he collects the bricks.

It has been said by one writer that I hypostatize time (i.e., treat it as something existing in its own right). I do not think there is the smallest justification for that charge, but—would it matter if there were? Modern science hypostatizes space.

* * *

In deference to Professor C. D. Broad I have excised from the present 1942 reprint the words 'at infinity', wherever these appeared in the previous issues of the book. It was obvious, from the description of the series, that the expression complained of was intended merely as an abbreviation of the cumbersome phrase '*the point where you decide to bring to a halt your endless chase after an observer who is not himself observed by one still more remote.*' In any case, no argument in the book was based upon the use of the words 'at infinity', so objections to the employment of that expression are not objections to the theory of Serialism. Perhaps an analogous example will make this more clear. I was taught, when young, to say that parallel lines met 'at a point in infinity'. Professor Broad would complain that such a definition was flatly self-contradictory. The question there is the very abstruse one of the meaning of the word 'infinity', and Broad would be entitled to his opinion. But suppose that he were to continue, 'Therefore the whole theory of parallel lines is founded upon a fallacy. Parallel lines cannot exist and we should be well advised not to travel by train!' The absurdity is obvious. Yet it is precisely a *non sequitur* of that description which he employs as his major weapon in all his numerous assaults upon Serialism.

The main opposition to Serialism comes, naturally, from those particular philosophers who were attacked so cautiously on pages 133 and 134 of this book. They belong to the group known as 'Ontologists', and their aim is to state what things actually 'are'. This may sound absurd; but, as a matter of fact, ontologists have one great achievement to their credit. They perceived the distinction between 'being' and 'existence'—a distinction which leads straight into the regress of Serialism. (That, probably, is why they mention it so seldom.) Now, these men were wont to claim, for the human mind, omniscience.

To admit the possibility of limits to human understanding would be, they held, to adopt a defeatist attitude detrimental to philosophical energy in the face of difficult problems. This purely politic decision hardened slowly into a rule (they loved rules). Anything which denied the omniscience of mind was to be regarded as involving a hidden fallacy. Thus, incredible though it may seem, they passed a law to forbid themselves from discovering a possible distasteful truth. It is not surprising, then, that the regressions of self-consciousness and of time, which issue the prohibited warning in the most emphatic fashion, were turned down without examination as certain to contain some flaw which it would be needlessly troublesome to locate. I pointed out that, in adopting this labour-saving attitude, they had been neglecting their job. Naturally, they wish now to show that they were justified in so doing. But can neglect be justified?

What was good enough to pass for an excuse in Ontology was advanced by Bradley in the old days. He said: 'Reality cannot be an Infinite Regress.' Note that this is a dogmatic assertion of what *is not*; and so, right or wrong, it is purely ontological. It does not introduce the question of man's mental capacities—a question which, since then, has been shifted, definitely, to another and more virile branch of philosophy, namely, Epistemology.

Serialism is not an Ontology.

There is a certain class of objects of knowledge which philosophy in general, ignoring the Ontologists, regards as 'given'. These are such 'Absolutes' as Time or Space or Sense-data or, we may add now, Self-consciousness. That these are 'given' means that they cannot be derived by logical processes from other items of knowledge, and that it is not possible to explain how we become aware of them. Our knowledge thereof is, therefore, not logical. It is with these Absolutes that Serialism is concerned.

Consider, to begin with, the absolute 'self'. Self-consciousness, in the sense of a Jones aware of Jones, is irrational. (If the reader wishes to be convinced of this, he should ask a physicist to expound to him why it is that a body cannot react to itself.) When, however, we introduce the notion of an absolute physical world known to Jones but other than Jones, we change irrational self-consciousness into something rational. Jones as known becomes a physical entity in that external world. This physical Jones known has (*vide The Serial Universe*) fewer characteristics than has Jones the knower; consequently, there is nothing irrational in Jones's awareness of this subordinate entity. There follows an infinite regress of Jones the irrational, self-conscious creature: he is replaced by an endless series of rational knowers each of which is aware of an external physical entity which serves him as an instrument and which he dubs, in popular phraseology, his 'self'. Such Absolutes as the sense-data of colour, sound and the like regress with the irrational Jones; for no merely physical Jones can be aware of these. Time, to take another Absolute, is, to a purely physical mechanism like Jones known, a standard distance (usually angular) traversed by a moving pointer. But it is essential that the pointer should be moving 'uniformly'. 'Uniformly' means here, traversing equal distances in equal seconds of absolute time. Now, it is Jones 2 who has to judge whether Jones 1's clock is moving uniformly (*vide*, again, *The Serial Universe*), so the knowledge of absolute time is transferred to Jones 2. Thence it is thrust off on to Jones 3, and so on as you chase the responsibility along the regress. The irrational knowledge of this Absolute regresses, consequently, with the irrational 'self-conscious' creature, leaving physical time as a standard space moved over with a standard velocity accepted arbitrarily as uniform.

But, halt where you choose in the regress (after the first term), you will have on your hands an observer with irrational 'self-consciousness' additional to his rational awareness of a subordinate physical observer (he knows that this subordinate 'self' is *his*) and with an irrational knowledge of Absolutes. It is the external world which becomes rational.

Now, this is not an ontological statement. It does not declare that there 'is' an infinite regress of observer or of time or of any sense-datum. The ontological statement is that Mind is irrationally aware of irrational Absolutes like sense-data or time or (though this must come into another book) space, and is even, most irrationally, aware of itself. The Ontologist has been faced with that statement for two thousand years, and he is naturally, at perfect liberty to continue wondering, for another two thousand, what he is going to do about it.

The serialist statement will be that *the serial process of extracting rationality from these irrationals is the process which produces the reliable world of Physics. That is a perfectly legitimate philosophy, and I believe that it must, in time, supersede all others.*

* * *

Mr H. G. Wells laments that I have taken something which he never intended to be treated seriously, namely, his description of 'duration as a dimension of space', and have brooded too much upon it. But it was not Wells's *Time Machine* which provided my starting point. (I used to argue with my *nurse* about the regress involved in her explanation of time as a length travelled-over by a 'now'.) What set me going was the impossibility of getting away from the popular notion, accepted emphatically by Newton, of Time as a *flow*—that is, as length passing a point. Whether the 'length' were simply spatial, or a dimension of extension more fundamental than either time or space, made no difference to the fact that 'flowing' Time was a regressive conception. I took Hinton's clever analysis, published ten years before Wells's joke, as a starting point, and set to work to discover what it was that he, and Newton before him, and generations of still earlier humans, were missing out—so that their descriptions seemed nonsensical.

It is always legitimate to indicate the order of successive states by a line equipped with an arrow. What required to be proved was the necessity of contemplating the travel of a 'now' along a line in which succession was already indicated; for it would be this travel which would make that line represent one-way space. Now, my forerunners had begun, in effect, by saying: 'Let there be a world of four dimensions and a three-dimensional observer travelling therein'—which begged the whole question. I, trying to get nearer to the epistemological standpoint, began: 'Let there be an observer mapping out in succession the states of an external world.' The result was that he would not only place those states in order of succession along a line, but would discover, in the process of construction, a subordinate 'self' moving along that line. He would bring to light the fact that the succession indicated by the mere line alone was succession for the *incomplete*, *external* world only—the world *minus* the map-maker. Succession in simple line fashion (without a 'now') for the *larger* world which included the discovered 'self' would require an additional dimension.

That secret, I think, could not have been revealed by the ontological approach. It was the reward of sound methodology.

Wells, after describing time-travelling as all nonsense, goes on to proffer an explanation of previsional dreaming —an explanation in which attention, occupying four dimensions, expands and contracts as it advances along time! There is nothing in this which had not been asserted in the present book (*vide* pages 161 and 162). But *I* said that this four-dimensional attention was the attention of Observer 2 following Observer 1: Wells wants me to alter this and to make it the attention of Observer 1. Why? It would be thoroughly bad methodology. It would not avoid the regress, with its consequent Observer 2, and so would be redundant. It would introduce absurdities; for, while this Observer 1 was expanded thus over a couple of days, another Observer 1 contracted to less than a second might blow the middle out of him with an electric spark. And it would *not* account for previsional dreams. According to Wells: the expansion of the time-travelling attention during sleep involves unconsciousness; its contraction involves waking. But to get a previsional dream of any clarity at all—a dream in which people move almost as in waking life—the contraction, and consequent waking, would need to occur at a point several days ahead of other waking people. None of these difficulties apply to Observer 2.

In brief, Wells's suggestion is a very valuable illustration, for the youth of this generation, of the kind of liberty which Victorian Materialism granted itself in the presentation of its chaotic case.

Wells has complained of my describing English materialism as insisting upon the 'eternal extinction' of every individual. He declares that this is a 'question-begging phrase', and says that it 'shows the quality of my thought'. This surprises me considerably. I should have supposed that Wells would have heard of Nietzsche, the arch-apostle of that brand of materialism which permeated German Court and military circles before the war of 1914. Nietzsche believed in an extinction which was not eternal. The entire world, according to him, was a mechanical dashing about of material atoms forming endless combinations. In course of time, he held, any given combination was bound to repeat itself; consequently each of us, after aeons of extinction, would find himself repeating his past life. The theory was glib nonsense. Nietzsche was completely ignorant of science and did not know that the Second Law of Thermodynamics expressly forbids any such repetition of a past state of the material world.

APPENDIX
TO THE THIRD EDITION

★

I. A NOTE BY
SIR ARTHUR EDDINGTON, F.R.S.

This extract is printed by permission from a letter written by Professor Sir Arthur Eddington. ('Minkowski's world', referred to therein, is the 'space-time' world adopted by Einstein for the purpose of his theory.)

'I agree with you about "serialism"; the "going on of time" is not in Minkowski's world as it stands. My own feeling is that the "becoming" is really there in the physical world,[1] but is not formulated in the description of it in classical physics (and is, in fact, useless to a scheme of laws which is fully deterministic).

Yours truly,

A. S. EDDINGTON.

Observatory,
Cambridge,
1928, Feb. 1.'

[1] AUTHOR'S NOTE.—This, I think, no Serialist can deny. The inclined line $O' O''$ in Fig. 9 is, clearly, as objective to the observer as are any of the vertical lines in that diagram. The fact seems to be that what is abstract to the first-term observer is concrete to him in his second-term outlook.

II. THE AGE FACTOR

If the dream theory propounded in this book is true, the extent of Time 1 future open to a dreamer's exploration grows smaller and smaller as that person's 'now' travels from the birth point to the death point in his body's history. The past part of the field, of course, grows correspondingly larger. These variations are exhibited very simply in Fig. 14, where the vertical line drawn through each age point is divided by the diagonal into

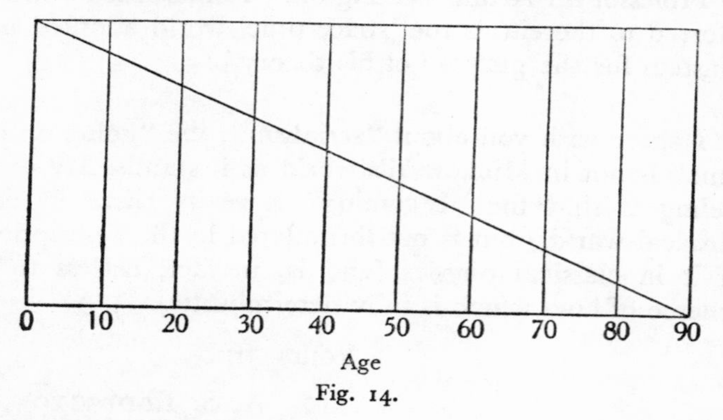

Fig. 14.

facilities for precognition (the lower part) and facilities for retrospection (the upper part).

This figure, however, is of little practical value to the experimenter. It assumes that the dreamer's attention wanders freely throughout the entire range of the field, so that the precognitive and retrospective elements in the dream might relate to waking incidents many years ahead of or behind the 'now'. The experimenter would have to wait until the end of his life in order to test the validity of the diagram.

Suppose, now, that we were to limit to one fortnight the period during which we seek, in waking life, for fulfilment of prophetic elements in the dream. Let us call precognitive elements which refer to that period alone, 'significant' precognitions. Then the facilities for such significant precognitions would be represented by little more than the bare thickness of the diagonal line, while the facilities for retrospection would be indicated still by the lengths of the verticals above the diagonal. The verticals below that inclined line would represent facilities for precognitions of no value in the experiment. Obviously, the facilities for retrospection in any dream would be enormously greater than the facilities for significant precognition.

Suppose, next, that we limit also to one fortnight the past period of waking life in which we look for evidence that the dream-images have been retrospective. Retrospection within those limits we will call 'significant' retrospection. Then the facilities for significant precognition and significant retrospection will be exactly equal—except, of course, in the fortnights succeeding birth or preceding death, which periods we must suppose to be left out of the diagram. That diagram, representing the comparative facilities at different ages, will take the form, now, of Fig. 15.

The lengths of the verticals below and above the medial line indicate the proportions between facilities respectively for significant precognition and significant retrospection; it being assumed that attention wanders quite freely over the fortnightly periods on each side of the 'now'.

But, even when we regard attention as wandering without aim, we cannot assume that its *concentrations* are entirely uncontrolled. Pending evidence to the contrary, we must suppose that the basic laws of psychology hold good; and one of those laws is that there is no concentration of attention unless *interest* is aroused. An equally irrefragable rule is that unless there is a concentration of attention no memory image is formed. Now, our experiments can deal only with such dreams as are *remembered*. And, before we can convert Fig. 14 into a diagram representing the probable comparative proportions of significant precognitive and retrospective

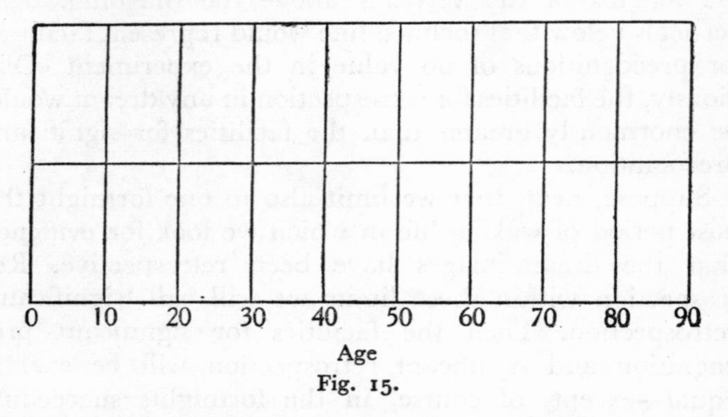

Fig. 15.

elements in a *remembered* dream, we shall have to tackle the question of the trend of *interest* at different ages.

All discussion of that problem must be highly speculative. Nevertheless, there are one or two broad generalizations which may serve us as rough guides in this very hazardous investigation.

Let us begin by considering the simplest form of interest, *viz.*, interest in what is *new*. The attention of a very young child would be arrested, probably, more by the familiar and comprehensible than by the strange and unintelligible. And, at the other end of the scale, we have the generally accepted fact that the interests of elderly people lie almost wholly in the well-known past. The attention of an old man, confronted with the network of associational tracks leading some to the future and some to the past, would be likely to take the backward path and remain happily among the scenes and experiences of bygone days. Somewhere between these two eras in a man's life would lie the period where interest in the new is at its maximum. We may follow popular opinion so far as to place this peak on the youth side of middle age.

But novelty is not the only exciter of interest; probably, indeed, it is not nearly so powerful a stimulant as is *desire*. And we have to recognize that (according to psycho-analytical theories) a great many dreams are deliberately framed structures giving expression to the desires of the dreamer for a world less disappointing than the one which he has experienced in the past. This dream-building demands some control of the bricks—which is, ultimately, control of the movements of attention—and the power of that control appears to increase with practice, just as in waking life. Now, the frustrations which the dreamer seeks to rectify are matters of *past* experience. In the past, also, lie the brain images which became associated with those frustrations—the so-called 'symbols' of the psycho-analyst. It is to the past, therefore, that, in the majority of cases, the constructor of a 'wish' dream will turn. That, of course, cannot be an invariable rule: the denizen of a mean street, starved for beauty, yet with beauty lying in the lap of Fortune ahead, would obtain his or her 'wish-fulfilment' most easily by a simple process of *forward* travel in the dream. Many parallel examples will occur, probably, to the reader. I should hazard, however, that most wish-fulfilment dreams demand retrospection, and that this is almost invariably the case where elderly people are concerned.

Now, let us represent by the number 2 the total amount of interest arousable in a dreamer at a given instant of his life, *i.e.*, at a given position of his travelling 'now'. And let us regard this total interest as distributed between past and future in proportions to be ascertained. If the interest is wholly in the past, there will be *no* interest in the future: conversely, if there is no interest in the past, the amount of interest in the future will be equal to 2. Intermediate numbers may be regarded similarly as values attributed to interest in the future.

The amount of interest in the future, treated thus as varying between 0 and 2, will become a factor by which we must multiply the chances of significant precognition

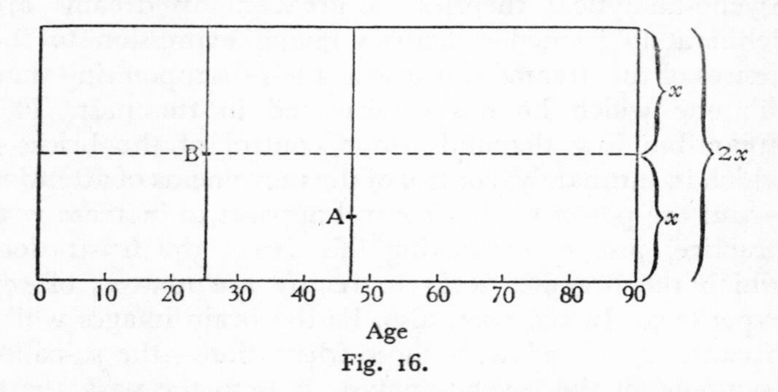

Fig. 16.

at the instant in question as indicated in Fig. 15. And the resulting product, which may fall above or below the medial line in Fig. 15, will indicate, by its division of the vertical line into two parts, the comparative chances of significant precognition and retrospection in a dream occurring at that moment. For instance, in Fig. 16, the point *A* divides the vertical line at the 47 age into lengths having the values of $\cdot 5x$ for the lower part and $1 \cdot 5x$ for the upper portion. Here, interest in the future has been given the value ·5, so that interest in the past becomes 1·5. The previously determined chances for significant precognition or retrospection were equal, each possessing the same value x. The two products $x \times \cdot 5$ and $x \times 1 \cdot 5$ are the lengths divided off on the vertical. The point B represents an imagined distribution of chances at another age. Here interest is given, again, the total value of 2. For it is quite immaterial whether the total interest aroused here is greater or less than that excited when the 'now' is at A; it is the way in which whatever interest there may be is *distributed* which gives the chances of the movements of attention.

Now, in view of the highly speculative character of the Interest factor, we cannot hope to draw the correct

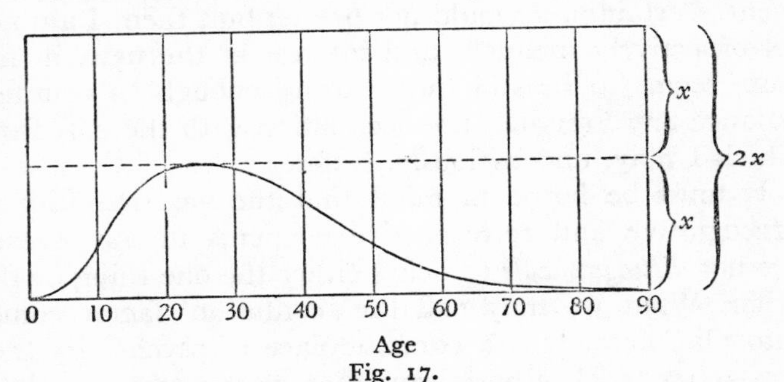

Fig. 17.

curve showing the comparative chances of significant precognition and retrospection at different ages. But, if the broad generalizations in which we indulged a little while ago are not entirely erroneous, the curve must have some such general *shape* as that shown in Fig. 17. The apex may be higher or lower, and it may be nearer to or farther from middle age; the beginning and end of the curve may not fall quite so low as 0; but the acceptance of a crest on the youth side and a trough on the age side seems to be our safest policy until we can get the curve determined properly by experiment.

I have given to interest in the future at the age of 25 the value of 1. (Interest in the past would have, in that case, an equal value.) There is some experimental justification for this distribution at that place (though, of course, a great deal more is needed), so I have used that evidence to provide a starting point for a curve which, otherwise, would never get started at all. Also, I have treated that point as the highest in the curve. For this there is little justification beyond commonplace caution.

The reader will realize, I assume, that the diagram is a guess at an average curve for a very large number of persons. It makes no allowance for unusual temperament. Certainly, it would not fit me; but, then, I am not retrospectively inclined, and interest in the new, in my case, seems to be still quite strong enough to counterbalance any Freudian pre-occupation with the past from which I may, unconsciously, suffer.

It must be borne in mind that the vast majority of precognitive and retrospective elements in any dream are not *distinguishable* as being either the one thing or the other. What we may call the 'confusion' factor comes into play here. It is a commonplace of psychology (not a peculiar tenet of Serialism) that most structures of the imagination are 'integrations'—blends of several images associated with several different waking impressions. And it is accepted generally that dream-images are mostly of the same character—certainly, it is very rarely that one comes upon one of these exhibiting an unmixed, photographic resemblance to any scene of waking life. Now, the possibility of discovering in one of these composite structures an element distinctive enough to be recognizable as pertaining to a chronologically definite incident of waking life depends, mainly, upon what may be described as the *coarseness* of the blending. The more intricate—the more fine-grained—is the integration, the more difficult becomes its analysis. And, with practice in constructing dream-images, just as with practice in waking imagery, the integrations do become more fine-grained, more beautifully blended, and, so, less easy to associate with any chronologically distinctive waking incident, past or future. But increasing practice means increasing age.

This Confusion factor has no effect upon the ratio of recognizable precognitive elements to recognizable retrospective elements at different ages; for it militates equally against the discovery of either. But the fact that it militates increasingly as the experimenter grows older means that, if this individual is seeking for evidence of precognition only (instead of trying to discover the ratio in question), his chances of success dwindle with advancing years. That applies, of course, to persons attempting the experiment described in the body of this book. Now, Fig. 17 shows already that, after middle-age, there is a rapid falling-off in the chances of the *occurrence* of precognition, recognizable or otherwise. If we take the Confusion factor into account, that falling-off must become still more pronounced.

The moral is obvious. The man who wishes to obtain evidence of his own precognition would be well advised to make his experiment when young. The longer he delays, the smaller may become his chances of success. To readers who already are elderly I must present my apologies for not having written this book sooner. Their best policy, it seems to me, is to conduct an experiment by proxy, under their own control. But in this connection a warning is necessary. The experiment is immensely fatiguing, and no immature person should be asked to experiment for more than two nights in succession, with a week's rest before the next attempt. In the cases of children, indeed, it would be far better to get the required information from a mass experiment as described in Appendix III, for then no single child need be asked to make more than one effort to recall its dreams.

III. THE NEW EXPERIMENT

The main drawback to the experiment dealt with in Part III of this book is that it demands a great deal of time at the one period of the day when nobody has any time to spare. To recall and write down with proper detail the dreams which have occurred just before waking takes from twenty-five to forty minutes. Consequently, unless one can afford to be late for breakfast, one needs to be called forty minutes earlier than usual—which is not, in most households, an easy matter to arrange. The only persons who are really free to write down their dreams in unhurried fashion are those who are sufficiently their own masters to get up when they please. Unfortunately, few people of the age we want to test (eighteen to thirty-four) fall within that category; not even when they are (as they should be in this experiment) holiday making in new scenes. Personally, I can get over this trouble by the fairly simple process of going to bed an hour earlier, which results in my waking proportionately sooner; but, for most, the only solution lies in the purchase of an alarm clock.

The next trouble lies in the extremely tiring nature of the experiment itself. Everyone seems to be agreed about this. Recalling one's dreams induces very great mental fatigue. Moreover, the previous determination to remember those dreams begins, after the fourth or fifth day, to affect one during the dream itself. One realizes, actually, that one is dreaming and that one must make an effort to fix the dream in one's memory. The resulting worry is detrimental to sound sleep, and people stop the experiment for fear of inducing insomnia.

The net result of these two difficulties is that the majority of experimenters, starting with the utmost enthusiasm, drop out after the third day; while those who persist longer are apt to declare themselves satisfied as soon as they have obtained the first minor result.

Clearly, then, it is advisable to devise some means of cutting down the number of nights required of each experimenter. And here another difficulty arises. It takes the average man at least three days before he learns to write down all the minor details of any dream episode which is unusual, and to avoid that labour when the incident in question is too commonplace to be of evidential value even if 'fulfilled' next day. Certainly, it will take him three days to acquire a glimmering of the art (for it is an art) of *noticing* the connection between his dreams and the corresponding incidents of waking life, past or future. In fact, the first three days' work is merely training for an experiment which does not begin properly until the fourth day.

Again, any reduction of the number of days, cuts away the foundation of the old experiment. There, the object is to ascertain what *proportion of persons* can perceive, in the course of fourteen days, effects strong enough to give rise to a suggestion of precognition. The proportion of persons who can discover such effects in the course of only three days will be, obviously, considerably smaller, and it will be reduced a great deal further by the fact that the first three records are those which have little more than an instructional value. The result will be, in all probability, to leave the question of whether such persons are to be regarded as normal or abnormal still undecided.

Suppose, now, that we were to get, say, 1,400 people to experiment for 3 nights apiece, producing 4,200 records. Can we assert that this would be equivalent to 300 persons experimenting for 14 nights each? No, this method of calculation would give us no information bearing upon our question, *viz.*: What is *the distribution of the observed effects among individuals*? For example, suppose that 200 instances of the effect were observed. This might mean that 1 person in 7 had been successful—a result which, in the peculiar circumstances (only 3 nights' experimenting apiece), would be highly favourable to the theory of normality. But it might mean, equally well, that roughly 1 person in 21 had observed 3 effects each, and then it would be extremely difficult to account for the fact that the remainder had observed nothing at all.

Still, it may be asked, would not the result of dividing the total number of dreams recorded by the total number of effects observed tell us how many records the *average* man would require to make before he could expect to observe anything worth noting? No—and this for two reasons. In the first place, the best results (those obtained between the eighth and fourteenth consecutive days) would be missing in this mass experiment where nobody records for more than three days. And, even if this were not the case, even if the first three experiments made by every individual were as skilfully conducted as would be later and more experienced efforts, the average number of dreams per effect in this mass experiment is not the same thing as the proportion of dreams per effect to be expected by the *average man*. For the existence or non-existence of this 'average' observer of effects is precisely the point which is in dispute and which the experiments have to settle. As we saw in the example a little way back, the same average of effects observed to dreams recorded can be obtained with considerable variation in the proportion of successful to unsuccessful experimenters.

I am going to show now that the proportion of total effects observed to total dreams recorded in any such mass experiment as we are considering is of no significance whatsoever. What *is* of importance—of decisive importance—is *the proportion of effects suggesting precognition to similar effects suggesting retrospection.* And, when we have realized this, we shall perceive the possibility of a new and really scientific experiment.

* * *

Imagine that a very large number of dreams have been examined for incidents resembling waking events, the latter being sought for within equal periods (preferably short) before and after the dreams. It is essential, of course, that the waking events in question should be chronologically distinctive—a dream incident resembling waking events which have occurred *both* before and after the dream is not to be counted.

The expressions 'resemblances to the past' and 'resemblances to the future' will recur so frequently in the coming argument that it is advisable to abbreviate these in some fashion. We shall call the resemblances to the past 'P-resemblances', and the resemblances to the future 'F-resemblances'.

Any resemblance, again, which is due to the dream being a result of the waking experience we shall designate, briefly, a 'causal resemblance'.

Now, we have to realize (what is often overlooked) that among the P-resemblances there are likely to be many which are *not* causal resemblances, but are due to pure coincidence. And the probable proportion of these is a thing which can be calculated. Suppose that we find that the chances of a particular coincidental P-resemblance are one in ten—which is written 1/10. Now, if we have discovered among the P-resemblances in a large collection of dreams one thousand of this value (*i.e.*, with the chances of coincidence equal to 1/10 in each case), we may say that one-tenth of those thousand cases were probably pure coincidences, the remaining nine-tenths being causal resemblances.

Noting this, we turn, let us suppose, to the F-resemblances and discover there one hundred examples of a similar kind. Are we to say, in this case also, that one-tenth of these were probably coincidences and the remaining nine-tenths causal resemblances? If we were to commit ourselves to any such supposition, we should be absolutely and entirely wrong. The evidence—the *total* evidence—would suggest that the whole hundred of the F-resemblances were pure coincidences.

For the chances that a purely coincidental resemblance to a dream will be discovered within some given period of waking life are entirely unaffected by the question of whether that period is before or after the dream. Consequently, in the case we are considering, we should expect to discover among the F-resemblances a number of coincidental resemblances *equal to the number of similar coincidental resemblances which have occurred among the P-resemblances*. Of the one thousand discovered P-resemblances, we saw that one hundred were probably coincidences. We should expect, therefore, by the ordinary laws of chance, to discover one hundred F-resemblances of similar value; and this is precisely the number we have found. There are not, then, the smallest grounds for supposing that those F-resemblances involved anything beyond expectable coincidence. The results we have imagined, therefore, would support the old theory of dreams.

If the 'probability' of each of the individual P-resemblances (*i.e.*, the chances of its being due to mere coincidence) were 1/100 instead of 1/10, and if the total number of these discovered were 500, then one in every hundred, which is 5, would be (probably) coincidences, and the remaining 495 would be causal resemblances. We should expect then to discover 5 resemblances of this probability (1/100) among the F-resemblances.

In short, if $1/a$ is the probability value of the kind of resemblance we are considering, and if n is the number of P-resemblances of this value, and N is the number of F-resemblances of this value, the laws of chance assert that $N = \text{(probably) } n \times 1/a$, *e.g.*, in the case just considered, $N = \text{(probably) } 500 \times 1/100 = 5$; if the old theory of dreams is right.

When the number of instances considered is very large, predictions and equations based on the laws of chance approximate very closely to those based on exact science. So, in such a case, we can say absolutely that, if the old theory of dreams is right,

$$N = n \times \frac{1}{a}.$$

We can, of course, read this as

$$n = aN$$

which means that, if we discover N F-resemblances of values $1/a$, there should be a times that number of similar resemblances among the P-resemblances. To obtain a concrete example, we can reverse our previous illustration. If we discover 100 F-resemblances with probability values of 1/10 in each case (so that $a = 10$), we should expect to find 10 times that number, each of similar probability value, among the P-resemblances, *i.e.*,

$$n = 10 \times 100 = 1000.$$

And, if we did discover these, the experiment would have favoured the old theory of Time.

But, if the number *n* of such P-resemblances turned out to be *less* than *aN*, then the evidence would suggest that some of the F-resemblances were causal relations. That suggestion would become satisfactory scientific proof of precognition, if the proportion of *expected* P-resemblances to *discovered* P-resemblances was very large.

This experiment has the following advantages.

(1) The number of records made is immaterial and may vary with different individuals takin part in the same experiment.

(2) The period, divided equally into past and future, in which results are sought for, may vary with the different individuals.

* * *

In the course of an argument with some 'Supernormalist' friends of the Society for Psychical Research, I raised the question of the effects of the Age factor, pointing out that these should render youth the ideal period for experiment. To illustrate my point, I arranged for experiments to be carried out by 22 volunteers from the University of Oxford, each of whom was required to make 21 records. To satisfy S.P.R. demands, I arranged that these records should be sent immediately after completion to Mr H. H. Price, Fellow of St John's College and Lecturer in Psychology, who forwarded them immediately, unopened, to the London Office of the Society. The occurrence of the possibly confirmatory future waking event was testified to, in each important case, by an independent witness. I should like to take this opportunity of expressing my thanks to all who assisted in this experiment.

The experiment was intended to be the old one, *i.e.*, it was designed to ascertain what proportion of the persons engaged could discover effects similar to those described in Chapter XI of this book—it being understood that each person should have completed 14 records. (21 were asked for merely to be on the safe side.) Unfortunately, the period chosen was a bad one—a little before the examinations known as 'Schools'. Consequently, most of the experimenters dropped out early, and only two completed the required 14 records. Of these two, one was a failure and the other a startling success.

I discovered at this moment, that one of my opponents was labouring under a remarkable misconception. He supposed that all the dream incidents which had not resembled clearly some waking incident of the near future should be counted as having been dreams of the past, so that the proportion of dreams resembling the future to dreams regarded thus conveniently as retrospective would be small in the extreme. His idea appeared to be to employ a computation of this kind as an argument against the new theory! I wrote to him, at once, pointing out that the only ratio of this sort which has any evidential significance is the ratio of dreams distinctively resembling the future to dreams similarly resembling the past when both are collected under similar conditions. And (judging this to be a case where one must employ a sledge hammer to crush a nut) I wrote to the Oxford experimenters asking them to examine their records for resemblances to the past within a period equal to that in which they had sought for resemblances to the future. Six sent me carefully compiled analyses on these lines. The results are given below. To these I add (for the satisfaction of the S.P.R.) the results of a similar experiment by myself, the idea being to ascertain whether I should or should not be classified as a 'supernormal' individual.

This, then, was the first 'try-out' of what I have called the 'New Experiment'. As I have explained before, I have not the smallest intention of *basing* upon the mere evidence of dreams a theory of such consequence as Serialism. But, for the pure, empirical psychologist, the ratio of 'Good' P-resemblances to similar F-resemblances apparent in the following seven analyses provides, it seems to me, statistical evidence of precognition amounting to practical certainty. And it is only the fact that every scientific experiment requires repeating several times which prevents me from labelling these results as scientifically *conclusive*. The departure from what would be, on the old theory of Time, the ideal ratio, is too enormous to allow us to entertain the idea that further extension of the experiment could have brought about a recovery.

The valuation of the results was made by myself a considerable time after these were in my hands. Illness prevented me then from instituting all the enquiries by letter which would have been necessary to ascertain the precise probabilities in each case; but this, as it turned out, was unnecessary. For the probabilities in the best class F-resemblances are certainly of an order of magnitude which is less than 1/10,000. And, on the older theories, each of these would require to be balanced by 10,000 *similar resemblances on the past side.* (When the reader realizes *that*, he will appreciate the potency of this new kind of experiment.) Consequently, I have adhered, below, to my original rough method of classification, *i.e.*, as 'Good', 'Moderate' or 'Indifferent'.

I found, however, that, in this business of judging the probabilities, there is a psychological trap of unexpected potency. One knows that the chances of a coincidental resemblance are quite unaffected by the temporal position of the waking incident—*i.e.*, whether it comes before or after the dream. But to know this, and to put it into practice, are two very different matters. The knowledge that a causal connection is also possible in the case of the P-resemblances affects one's judgement of the *entirely independent* chances of coincidence. Let me give the reader an example. One of the subjects dreamed of a head divided up into sections 'like a phrenologist's chart'. He remembered having seen such a head, a month before, in a medical book belonging to a friend at the university. That looks like a clear case of retrospection, does it not? Well, I have practised a deception—the book was seen *after* the dream. What sort of evidence of precognition is that? What are the chances of the subject coming across such a picture in the space of one month when he is in the habit of visiting a friend who is studying medicine? One's judgement now is very different. Yet the chances of coincidence were precisely the same in each case. And, now that the reader realizes this, I will admit that the book was, after all, seen before the dream.

The only safe way of judging such resemblances is to proceed in complete ignorance as to whether the resemblance was to the past or to the future; and, in all experiments of this kind, the judge should be left in that position. I propose to do this for the present reader. I shall describe the dream and the waking event resembled without giving him the slightest hint as to whether that waking event was before or after the dream. He can decide, then, for himself, whether or not he agrees with my valuation. If he disagrees, he can note his own valuation in the margin—it will make little or no practical difference to the ultimate result. In the end, I shall tell him which dreams referred to the past and which to the future, but I strongly advise him to make his own estimate of values without looking ahead. I got over the difficulty myself by treating the whole lot as if they were resemblances to the future.

One difficulty arises which must be dealt with before we proceed further. A man dreaming of a past scene or person known to him writes down afterwards, 'Saw So-and-so', or, 'Was in such-and-such a place', and one must assume, without requiring further detail, that this was retrospection—if the scene or person is not seen also after the dream.

But, if one counts such undetailed resemblances to the past, one must accept equally the subject's judgement when he claims a resemblance to a person or scene observed after the dream. Otherwise, one would be cutting out a class of F-resemblances without making any corresponding reduction in similar resemblances to the past. It will not solve our difficulty if we insist that all claims to visual resemblance must be fully detailed. For none can write down *all* the detail in a dream. And the dreamer, recognizing a dream person as someone he has met in the past, would be aware, in the morning, that the dream was evidence of retrospection, and would fill in the necessary detail. He would have no such recognition to warn him if the resemblance were to some stranger to be seen after the dream. We have no choice, then, save to accept the dreamer's judgement in all claims to a merely visual resemblance, past or future, and to warn him to be exceptionally careful in making them. Fortunately, the evidence is seldom of this purely photographic kind—the important part of the dream is usually an *incident* in which the scenery plays merely a supplementary part. In the present analyses, for example, there are only two such claims unsupported by written or sketched detail.

The average age of the subjects was in the neighbourhood of twenty years. The period searched for waking incidents extended for two-and-a-half months both before and after the dream.

And now here are the results.

SUBJECT A

REMARK. This subject is an artist, and his resemblances are all of the purely visual kind. An unusual and interesting case.

1. This dream occurred before the period of the test.
Value. None.

2. The waking event was outside the time limit.
Value. None.

3. The dream was of a 'small curtained cell' connected with phrenology.
Waking event. The subject consulted a palmist in a tent which he describes as 'an almost exact replica' of the curtained cell.
Interval. Six weeks.
Value. Indifferent.

4. The dream was of a head divided up like a phrenologist's chart.
Waking event. The subject saw a head divided up in this fashion in a medical book belonging to a friend at the university.
Interval. About a month.
Value. Indifferent.

5. The waking event was outside the time limit.
Value. None.

6. The waking event was outside the time limit.

7. The dream record describes 'a canoe made of thin brown varnished wood—with a green canvas cover for a short piece at one end'. It continues, 'My sister and I were up at the green canvas end, and with great glee she fastened the canvas across over her head.'
Waking event. The subject writes that he took his small sister, who was at school at Exmouth, out for a ride in a speed boat there. They were provided with pieces of green waterproof sheeting to protect them from the spray. He adds, 'As we started off, my small sister was very excited and pulled the green waterproof sheet right up to her chin, and for a moment ducked her head under it.' He suggests that the visual picture offered by his excited sister 'sitting in the corner of a square

stern' with the green sheet up to her chin was the origin of the dream-image.

Interval. Two-and-a-half months.

Value. Indifferent (owing to the length of the interval).

8. The dream record runs: 'Dreamed I was sailing alone in a small boat constructed out of a number of cart-wheels cut in half with planks nailed along and tarred.' The record contains the sketch of this structure given below.

Waking event. The subject saw a wooden structure, of the same shape as that sketched in the record, being carried by a man. He enclosed a second sketch (of this real structure) which shows three half-circles of wood joined together by slats of wood, the whole being precisely similar to the dream-image except that the wheel *spokes* were missing.

Interval. Nearly a month.

Value. Good. The reader who wishes to estimate the probability of encountering such an object in the space of one month can proceed as follows. Let him ask his friends in succession whether they have ever seen such a structure, and let him continue the process until one answers in the affirmative. Then let him add together the ages, in months, of those who have replied in the negative. The total will be the denominator a in the probability fraction $1/a$. In my own investigation this fraction has reached, so far, to 1/5400; and, from my experience of engineering, I shall be surprised if the fraction comes out, eventually, higher than 1/10000.

9. The waking event was outside the time limit.
Value. None.

10. The waking event was outside the time limit.
Value. None.

11. The dream was of a fight with a private soldier.

Waking event. The soldier's face was that of a man met in waking life, and to whom the subject took a violent dislike.

Value. Moderate.

This subject completed six records.

SUBJECT B

This subject completed nine records, and noticed no resemblances to waking incidents either distinctively past or distinctively future.

SUBJECT C

12. The dream was of a car out of place in the traffic owing to children having meddled with it.
Waking event. A letter about a motor car accident.
Value. None. Details were quite different.

13. The subject refers to an alleged dream on the night of May 19–20; but I can find no such dream in his records.
Value. None.

14. The dream was that the subject was trying to take a photograph of a friend.
Waking event. The friend visited Oxford.
Interval. Apparently over a week.
Value. Moderate.

15. The dream was that the subject was offered some rolls to eat, and that these were not as hard as he liked them.
Waking event. Some rolls provided for the subject were not as hard as he likes them.
Interval. Several weeks.
Value. None.

This subject completed twelve records.

SUBJECT D

16. The dream was of a person lecturing. No details.
Waking event. A person lecturing.
Interval. One day.
Value. None. The subject pointed this out himself.

17. The dream was of a conversation about the Oxford 'Groups' movement. No details written.

Waking event. The occurrence of a conversation declared to be similar to the dream. No details given.

Interval. One day.

Value. None.

18. The dream was that a table in the subject's rooms had been moved (to the subject's annoyance) from its position next the wall into the middle of the room.

Waking event. An acquaintance entered the subject's room in the latter's absence and moved the table as described, in order to get a better light for something he was writing on that table.

Interval. One day.

Value. Moderate.

19. The dream was of the Vicar of St Aldate's leading an open air meeting.

Waking event. The subject saw the Vicar of St Aldate's leading an open air meeting.

Interval. Ten days.

Value. Moderate.

This subject completed seven records.

SUBJECT E

20. The dream record runs: 'Am wearing a dress with a white cowl-shaped collar of some artificial silk material—find it marked with something black, apparently soot. I try to get the marks out.'

Waking event. The subject writes: 'I washed some black stains from the white cuffs of a dress. The dress had a cowl-shaped collar. Both collar and cuffs were of artificial silk.'

Interval. Five or six weeks.

Value. Moderate. The subject possesses a dress with collar and cuffs of white artificial silk. What are the chances that she will get a black stain on one or the other in the course of five or six weeks and will try to wash it out? Note that it is the collar which is marked in the dream, and the cuffs in the waking incident.

21. The dream was that the subject was trying to work out some details of legislation in Roman history.

Waking event. The subject relates this dream to her first detailed study of Sulla's legislation.

Interval. Not stated.

Value. Moderate. I have accepted the subject's own verdict, mainly because the resemblance was (I may admit this here) to a past event. But the book was read, presumably, both before and *after* the dream, so that full details really are required before one can agree that the resemblance was to a waking incident which was *distinctively* past.

22. The subject dreamed that she was invited to meet a party of school children to be entertained for the day. Her friend and she try to entertain them in one room.

Waking event. The subject and friend help to entertain 'a number of visitors from a London settlement'.

Interval. I have omitted to note this; but my impression is that it was either one or two days.

Value. Moderate. The visitors were not children, and there was no mention of a London settlement in the dream. (The only place mentioned there was Worcester.)

23. The dream was of opening a parcel by slipping the string round the corners.

Waking event. She herself opened a parcel that way.

Interval. One day.

Value. None.

24. The dream record runs: 'Small boy says he is learning German by gramophone. Does not yet know any French.'

Waking event. It was suggested to a friend of the subject that she should learn German, and then that she should do some more French—in that order. The subject overheard this.

Interval. One day.

Value. None.

25. The record runs: 'R. F. consults me about some money. I go down to see her about it in the room below my own. The matter is not quite straight. She says she will see R. H. about it.'

Waking event. The subject relates this dream to the settling of some money question with the President of the Junior Common Room.

Interval. Three weeks.

Value. Indifferent. There must have been many other financial matters settled within the period of past and future allowed. The subject's dream record and commentary are too terse to be of much service here, and we have to trust to her judgement that the resemblance was in any way strong. Anyhow, the interval is a long one for a mere resemblance of uncertainty in money due.

26. The dream record runs: 'Go into College Hall and find my brother sitting there. Friend M. R. H. says he has altered since she saw him last.'

Waking event. The subject's brother visited her in College and dined in the Hall. Her friend made the remark as in the dream.

Interval. Three weeks.

Value. Moderate. The chances of her brother visiting her in College during an interval of three weeks are not strikingly small. And that her friend will remark that he has altered is highly probable—if he comes.

27. The subject dreamed of a 'seed' which takes the form of a 'honeysuckle flower . . . yellow with purple centre where the tips of the petals of each flower would be.'

Waking event. The subject reports that the first honeysuckle she saw that year had the same yellow and purplish flowers as the metamorphosed flower of the dream.

Interval. One day.

Value. Moderate.

28. In the dream the subject is seated at a table with the Principal of the College, and is eating soup flavoured with tomatoes. The Principal is wearing 'a grey dress of some soft material'.

Waking event. The subject reports that the Principal wore at dinner a grey georgette dress which she had not worn before. Tomatoes were served as a vegetable.

Interval. Ten days.

Value. Indifferent.

29. The dream record runs: 'As I woke I heard very clearly the phrase—*The Times*, dreams of four men.'

Waking event. The subject agrees that this dream was associated with her reading Mr John Buchan's book, *The Gap in the Curtain.* Mr Buchan wrote to me that this book had been inspired by *An Experiment with Time.* It is, essentially, the story of four men who practised the 'Dunne' experiment until they were able to foresee, conjointly, a page of *The Times* newspaper a year ahead. The interest lies mainly in the uses they made of this piece of foreknowledge.

Interval. About two months.

Value. Good. *The Times* has been in existence for over 100 years. And it is pretty safe to say that, during that period, there has never been, in literature or conversation, any specific association of that paper with 'the dreams of four men'—until Mr Buchan wrote his book. The chances that such an association would be made in print within the significant two months were, consequently, not more than 1/600, and probably a *very great deal smaller*. We have to multiply that figure (or the more likely smaller one) by the probability that the reference would be of such a character *as to come to the notice of the subject* within the required period. The final figure must be 1/several millions.

30. The dream record runs: 'Find myself in a small room as though hastily arranged for a meeting. I am sitting in the front row in a low chair. M. R. H. is on my left in a high one. There are railway posters on the left wall of the room. A man comes in to speak.... He says ... he has been to Chelsea on a holiday.... My mother gives me some tea knives to polish and put away.'

Waking event. The subject was present at a conversation which turned largely upon holidays. Later in the same conversation someone mentioned 'Chelsea'. (The subject says this last was unusual in her experience.) During the conversation she sat on the right of M. R. H., and on a chair much lower than hers. A large picture in the room was more than once referred to as a poster. Just before the conversation M. R. H. (apparently—the subject writes 'she') had borrowed two of the subject's tea knives. Afterwards the subject puts the knives which had been used together on a tray on the table.

Interval. One day.

Value. Good. It would be interesting to attempt an estimate of the probabilities here. For example: Tea party in friend's room, 1/2; sitting right of friend, 1/2; chair difference, 1/2; mention of 'posters', 1/100; mention of holidays, 1/1 (considering that the term was nearly over); mention of Chelsea (rare, according to the subject), 1/200; collecting tea knives, 1/2 (either she or her friend would have done it). Total $1/2 \times 1/2 \times 1/2 \times 1/100 \times 1/1 \times 1/200 \times 1/2 = 1/320{,}000$.

31. The dream record states that a conversation about fashions is occasioned by a picture of Lottie Lehmann.

Waking event. The subject saw the picture in some magazine.

Interval. Eight weeks.

Value. Moderate.

32. The dream record runs: 'I think I am going to Honolulu. I can see an island in the distance.'

Waking event. In a film which the subject saw, the approach to Honolulu was described.

Interval. Two days.

Value. Good.

33. The dream record states the subject and J. B. discuss a language which is being spoken. The subject says it is Norwegian or Finnish.

Waking event. The subject asked her friend if she had been working hard at her 'Norwegian grammar'—a slip of the tongue, as she had meant to say 'Old Norse'.

Interval. One day.

Value. None.

34. The dream was of a friend practising a peculiar gymnastic exercise known as 'prog'.

Waking event. Her friend demonstrated the exercise in question to her after bathing.

Interval. Two or three weeks.

Value. Good. Here, I am afraid, I must give the situation away. The resemblance was to a past event. But the word 'prog' did not, I gather, enter into the dream. It was used by the subject when awake and writing the record. The resemblance consists, therefore, solely in the details of the exercise and the fact that it was the subject's friend who performed it. The value, for example, would be reduced if the friend performed the exercise *after* as well as *before* the dream; for this would raise the question as to which waking scene the dream-image referred to.

35. The dream record runs: 'I look for letters. There are some folded forms in the pigeon-hole, a book of stamps, and a number of photographs. Coloured photographs are also pinned round the pigeon-holes. One of the forms is addressed to Miss Lee, and I wonder why it should be in the H pigeon-hole.'

Waking event. The subject writes: 'On . . . there were a number of notices round the pigeon-hole, one of them bright orange. It was a notice that unstamped letters should not be put in the letter box. I found a note addressed to Miss Richmond, and wondered . . . why it should be in the H pigeon-hole.'

Interval. One day.

Value. Moderate. Note the reference to stamps, and the connection (well known) between 'Lee' and 'Richmond'.

36. The dream record runs: 'Look out of a window and see a small boy dressed as a Red Indian. He is practically naked. . . . More small boys come to meet him—normally dressed. . . . I look into a garden and see a kind of bridge which they have made. . . . The hand-rail is formed of very long pieces, and is split in places. I lift up the separate pieces one by one.'

Waking event. The subject writes: 'I crossed a bridge over casmall creek. It was very roughly made of long and rather thin pieces of timber. It had a hand-rail on each side.' She goes on to note the resemblance between the real bridge and the dream hand-rail. 'The different pieces' (of the bridge) 'fitted very badly and moved under my feet as the rail in the dream did when I lifted the various pieces. Shortly after crossing the bridge I saw a number of small boys bathing.'

Interval. One day.

Value. Good.

37. The dream record runs: 'I am going to a concert at Balliol and wonder what to wear. Think I must put on a big coat if I wear a thin dress, as it is very cold.'

Waking event. The subject's friend (G. H. J.) discussed what she should wear for a concert. She said that earlier in the term she has always worn a fur coat to go to the Balliol concerts as it had been so cold.

Interval. One day.

Value. Indifferent.

38. The subject dreams that she is sitting on a low chair at a high gate-legged table. The chair stands in a patch of sunlight. A lady comes in and moves the table and 'the other chair' so that they stand in the sunlight too.

Waking event. The subject writes: 'I washed my hair and wished to sit in the sun to dry it. I moved first my table so that I could sit by it in the sun and then took a low chair over to the window and sat in it in a patch of sunlight.'

Interval. One day.

Value. Indifferent.

39. The dream record runs: 'My cousin J. H. has just become engaged. I see his fiancée for the first time, and am surprised that she has reddish hair.'

Waking event. The subject received a letter from home describing her cousin's fiancée and mentioning the colour of her hair.

Interval. Two or three days.

Value. Good.

40. The dream record runs: 'There are a number of people in a large building like a church. I suppose that they are prisoners.'

Waking event. During a conversation, 'the subject of Church services in prison was discussed'.

Interval. Two days.

Value. Moderate.

This subject completed 21 records.

SUBJECT F

This subject put forward three possible resemblances to the future, and two possible resemblances to the past. The resemblances to the future had no value. Of the resemblances to the past, one was outside the time limit, and the other was not chronologically distinctive.

This subject completed sixteen records.

SUBJECT G (myself)

41. The dream was that a Bishop had written to me suggesting that my wife and myself should be re-confirmed. I met him some ten days later.

Waking event. Was asked to tea to meet the Bishop who married my wife and me four years ago. It is a joke between my wife and myself that we should like to have our wedding over again; and I should, certainly, have mentioned this to the Bishop had I met him. Illness prevented me from going.

Interval. One to four days.

Value. Indifferent.

42. The dream was of a detective named 'Earheart'.

Waking event. A reference to Miss Earheart (the lady who flew the Atlantic) in the newspapers.

Interval. Under a week.

Value. Moderate. We have to consider simply the chances of coming upon such a name in the course of a week.

43. The dream record ran: 'A new Dunne-type aeroplane appeared in the sky, very noisy and climbing at a tremendous rate. . . . The effect was as if it were flying upside down. . . . Was told that it was one of a number of aeroplanes which were out searching for a missing airship.'

Waking event. The 'Dunne' aeroplane was the first of the tailless machines, and came to an end about 1916. In about 1924 this type was revived by Lieutenant Hill under the name of the 'Pterodactyl', built from my patents and with my consent and advisory assistance. I saw in a newspaper a picture of, apparently, a new 'Pterodactyl'; and, in another paper, I read that this machine, piloted by Lieutenant Stainforth (who held the world's speed record), was going to join with others in a 'balloon hunt' at Hendon. Later, I read an account of this 'balloon hunt'. The balloons were shaped like monsters (supposed to be invaders from Mars) and Stainforth beat the other defending craft engaged in chasing these monsters, and brought them down with shots from his revolver. He was flying a new 'Pterodactyl', and the newspaper accounts declared that he amazed the crowd by his evolutions around the 'monsters' as he brought them down.

Interval. Twenty and twenty-four days.

Value. Moderate.

44. The dream record runs: 'My father-in-law told me that he had had a dream of the future.'

Waking event. My father-in-law told me that he had just had the first dream of his life.

Interval. A week or ten days.

Value. Good. The chances can be calculated from the fact that it was the first remembered dream of my father-in-law's life.

45. In the dream, I was looking at the outside of a Cinema. The bill over the door showed that a film based upon a story by Owen Wister was being shown. The record states: 'This story was not *The Virginian*, but was a similar cowboy story about one of the Virginian's friends.'

Waking event. A friend who came to lunch began talking about revolver shooting, and, to illustrate a point, asked me if I had seen the film of Owen Wister's book, *The Virginian.*

Interval. Three days.

Value. Indifferent.

46. The record runs: 'My relative in America had broken his leg (later, his neck) by walking on a ledge which had given way.'

Waking event. I received a copy of a letter from the British Consul at Los Angeles stating that my relative was stranded there penniless and would, probably, be deported unless funds for his maintenance were sent. I spent most of the day writing letters about this affair. In the evening I read in a 'shocker' (*Tale of Two Murders*, by H. C. Asterley) of a man who had fallen from an ornamental ledge which ran along the side of a house. The book describes him (page 62) as 'lying there in the pathway with his left leg twisted up under his body in a sickening, horrible manner'; but the reader is left in doubt as to whether the man is dead. If he were so, it would mean that the hero's best friend had committed a murder (which I considered to be a very unlikely development). However, on the next page, it is stated that someone overcomes his reluctance to approach the man, and finds that he is, after all, dead.

Interval. Three days.

Value. Good. I have read shockers all my life, and have never before come upon a case which has deceived me into supposing that a character has only broken his leg, when, as a matter of fact, he has broken his neck. If I divided that period into three-day stretches, this gives me the probability figure 1/6083. That has to be multiplied by the probability of my receiving the letter from the British Consul during the significant three days. Personally, I should have put this last probability at 1/100.

47. The dream record runs: 'Was . . . looking at a small oblong building of dull grey brick. It bordered a deep gully

on the right as I faced it. The gully was crossed by a bridge on my right. There was a noise of trains.'

Waking event. Visited the 'Bluecoat School' (Christ's Hospital) at Horsham. Saw the bridge on my right crossing the gully, at Guildford, *en route.* Was lodged in the small house of dull grey brick (but with a red-tiled roof). It was at the bottom of another gully, and beside another bridge crossing this, but on the wrong side. A railway line passed twenty yards from the door; but I was told that no trains ran at night, so that I should not be disturbed. This last was a mistake. Trains thundered by apparently every half-hour during the night, and they woke me repeatedly.

Interval. One day.

Value. Indifferent.

48. The dream record states that on the small house mentioned in the last dream was an inscription saying that the building had been erected to 'the memory of Chevasse'.

Waking event. Saw a memorial to Bishop Chevasse unveiled.

Interval. One to two weeks.

Value. Good. The question is simply: What would be the chances of seeing, in 'one or two weeks', a memorial tablet containing the name Chevasse? The answer is easy. I, the person concerned, have seen such a tablet once, and once only, in the course of my life. If I divide that period into intervals of one-and-a-half weeks, the probability works out at 1/1800 (roughly).

I completed seventeen records.

The last dream ends the list of the results in this Oxford experiment. So, if the reader has completed his estimate of values, I will tell him which dreams appeared to refer to the future, and which to the past.

The resemblances to the future were dreams Nos. 3, 7, 8, 18, 19, 23, 25, 27, 28, 29[1], 30, 35, 36, 37, 38, 40, 41, 43, 45 and 46.

The resemblances to the past were dreams Nos. 4, 11, 14, 20, 21, 22, 26, 31, 32, 34, 39, 42, 44 and 48.

[1] This dream occurred before Mr. Buchan's book had been published or reviewed.

And here is the summary of results.

SUMMARY OF RESULTS

Subject	Number of records	Resemblances to the past			Resemblances to the future		
		Good	Moderate	Indifferent	Good	Moderate	Indifferent
A	6	0	1	1	1	0	2
B	9	0	0	0	0	0	0
C	12	0	1	0	0	0	0
D	7	0	0	0	0	2	0
E	21	3	5	0	3	3	4
F	16	0	0	0	0	0	0
Myself	17	2	1	0	1	1	3
Totals	88	5	8	1	5	6	9

CONCLUSIONS

It would be impossible, of course, to make these results square with any classical theory of dreams. For that, the P-resemblances would need to outnumber the above-recorded F-resemblances by many thousands. Nor is there the smallest hope that this enormous initial deficit of P-resemblances would be made good by extending the experiment further. The F-resemblances have far too large a start to give the P-resemblances any chance of overtaking them and getting ahead of them to the required extent.

It will be noticed that these Subjects, taken *en masse*, appeared to dream more of the future than of the past. It will be remembered that their average age (omitting myself) was in the neighbourhood of 21.

An important indication is one which does not leap immediately to the eye. There is no evidence anywhere of the existence of a special faculty for precognition. 'Supernormalists' may perceive that the F-resemblances of Subject E, 10 in all, outnumbered those of all the other Subjects put together. *But they should note, also, that her P-resemblances, totalling* 8, *outnumbered all the other P-resemblances recorded.* The evidence is simply that her dreams were more clearly related to distinctive episodes of waking life—past and future—than were those of the other experimenters.

My own case is, admittedly, unusual. I have never pretended that I am not an exceptionally good dreamer; and though I cannot, at 56, compete with Subject E in the twenties, I did, actually, beat the remainder. But here, again, if my F-resemblances numbered 5, my P-resemblances numbered 3. (Ignoring the indifferent results, my F-resemblances were 2 and my P-resemblances were 3.) There is no evidence that I possess a special faculty for precognition.

INDEX